FUNDAMENTALS OF
CIM
TECHNOLOGY

FUNDAMENTALS OF CIM TECHNOLOGY

DAVID L. GOETSCH

AUTOMATION IN DESIGN, DRAFTING, AND MANUFACTURING

DELMAR PUBLISHERS INC.®

NOTICE TO THE READER

Cover CIM wheel courtesy of Manufacturing Engineering SME's magazine *CIM Technology: CASA/SME's Magazine of Computers in Design and Manufacturing.*

Cover photos courtesy of Tandem Computers Incorporated

Delmar Staff

Associate Editor: Joan Gill   Production Editor: Christopher Chien
Assistant Managing Editor: Gerry East   Production Coordinator: Linda Helfrich

For information address Delmar Publishers Inc.,
2 Computer Drive West, Box 15-015,
Albany, New York 12212-5015

Copyright © 1988 by Delmar Publishers Inc.

Printed in the United States of America
Published simultaneously in Canada by Nelson Canada,
a division of International Thomson Limited

10 9 8 7 6 5 4 3 2 1

Library of Congress Cataloging in Publication Data

Goetsch, David L.
   Fundamentals of CIM technology: automation in design, drafting,
and manufacturing/by David L. Goetsch.
      p.   cm.
   Includes index.
   ISBN 0-8273-2844-3   ISBN 0-8273-2845-1 (instructor's guide)
   1. Computer integrated manufacturing.   2. CAD/CAM systems.
I. Title.
TS155.8.G62 1987   87-25959
670.42'7—dc19   CIP

CONTENTS

PREFACE

FUNDAMENTALS OF CIM TECHNOLOGY was written in response to the need for an easy-to-understand but comprehensive, up-to-date text on automation in design, drafting, and manufacturing. There are several books available for fourth-year engineering and graduate students. However, there is little available to meet the specific needs of students in postsecondary vocational schools, community colleges, and technical schools; nor is there a book which meets the special needs of the Baccalaureate student in technology.

WHO IS THIS BOOK FOR?

FUNDAMENTALS OF CIM TECHNOLOGY can be a useful text for a variety of categories of postsecondary students, including:

- Manufacturing technology students
- Industrial technology students
- Engineering technology students
- Drafting and design students
- Machinist students
- Technology education students
- Industrial arts education students
- Vocational teacher education students

Practitioners in design, drafting, and manufacturing who need technical updating in the areas of CAD, CAM, and CIM may also find this book useful.

SPECIAL FEATURES OF THE BOOK

FUNDAMENTALS OF CIM TECHNOLOGY has several special features that enhance its usefulness as a learning aid. The most important of these are:

1. The book is written in simple language making the technical concepts dealt with easier to understand.
2. Several chapters begin with "real-life" vignettes.
3. The book is heavily illustrated with a broad range of drawings and photographs which complement the text.
4. The applicable levels of math are covered in appropriate chapters.
5. Each chapter contains a summary, review questions, and problems.
6. The book contains a comprehensive glossary of CAD/CAM/CIM terms for reference.
7. The book contains a comprehensive index for easy access to specific material.

FUNDAMENTALS OF CIM TECHNOLOGY begins with an introduction that puts the many buzzwords associated with these concepts into perspective. Chapter 1 gives an in-depth explanation of the computer. This is important because the computer is at the heart of all CAD/CAM/CIM developments. Chapter 2 explains how the computer is actually used in CAD/CAM/CIM settings and what types of computers are used.

Chapters 3 through 7 are the technical chapters. Each is a mini-book covering a major CAD/CAM/CIM concept such as CADD, CNC, Robotics, group technology, and CIM/FMS. Chapter 8 is an in-depth treatment of CAD/CAM/CIM, management. This is an important chapter because many of the CAD/CAM/CIM students of today will be the CAD/CAM/CIM managers of tomorrow. Chapter 9 covers several other CAD/CAM/CIM concepts that are important, but less known than those presented in Chapters 3 through 7. Chapter 10 is a look at the future of CAD/CAM/CIM.

FUNDAMENTALS OF CIM TECHNOLOGY was purposely designed to have both broad appeal and usefulness. I hope this book meets the needs of a broad and varied audience in a language that is easy to understand.

ACKNOWLEDGEMENTS

The author wishes to thank the following people for their valuable assistance in reviewing this book. Through their input, the book has been made much better than it other wise might have been.

Larren Elliott
Texas A&M University
College Station, Texas

Wayne Kenner
Corning Community College
Corning, New York

Kent E. Kohkonen
Brigham Young University

Wallace Pelton
Texas State Technical Institute
Waco, Texas

Robert E. Speckert
Miami University
Hamilton, Ohio

INTRODUCTION

Manufacturing firms in this country and abroad are working toward the implementation of computer integrated manufacturing (CIM) in an effort to be competitive in the national and world markets now and in the future. Many have already automated some of their traditional manufacturing functions and processes by implementing computer-aided design and drafting (CADD), computer numerical control (CNC), robotics, and group technology. The next step is to integrate these and the various other processes used to convert raw materials into finished products. The result will be computer integrated manufacturing.

In order to understand CIM, one must examine the entire spectrum of development in manufacturing to see where it fits in with what has preceded it and what will come after it. Manufacturing has evolved over the years through several distinct phases. Each of these phases falls along a continuum. At one end of the continuum are man's earliest attempts at manufacturing. At the other end, one finds the completely automated manufacturing plant. In between lie the various stages of development which have occurred and will continue to occur until the goal of the automated factory is realized at some point in the future.

The earliest stage in the development of manufacturing can be called the MANUAL PHASE. In this phase, people used their hands and simple, crude implements to produce the products they needed to live. These products were such things as tools, weapons, furniture, and so on. In spite of the crudeness which characterized early manufacturing, there was one important advantage that was lost in the later stages of development and is only now, after hundreds of years, beginning to resurface. That advantage is called INTEGRATION.

The earliest manufacturing processes were completely integrated because one person did it all. An early craftsman designed a product in his mind and performed all of the processes required to produce it. Because the craftsman's mind and hands were connected, there was continuous and immediate feedback as he moved through the various processes in the making of the product. Because they were controlled by his mind, his hands would produce the product he saw in his mind. Such control, coordination, and continuous feedback are known as integration. Not since the days of the early craftsman has there been total integration.

The next stage in the development of manufacturing can be called the MECHANIZATION PHASE. It was brought about by the industrial revolution. In this phase, manufacturing tools and processes became mechanized and specialization became the norm. All of the various manufacturing processes were divided into categories such as casting, forging, turning, milling, drilling, and cutting; manufacturing workers became specialists in one of these areas.

Specialization resulted in the elimination of integration and the separation of design from manufacturing. It also created the need for drafting, since one person would now design a product and several other specialists would manufacture it. Drawings became the vehicle for communication between design and manufacturing. Specialization also created a need for quality control, since each specialist saw only his small part of the overall production process and performed only his specialized task.

The mechanization phase had the advantage of being able to accomplish mass production, to produce interchangeable parts, and to produce many more products faster. However, it lacked the old advantage of integration.

The current phase in the development of manufacturing is called the AUTOMATION PHASE. There have been several important milestones in this stage. In 1975, mass production was automated through the use of transfer lines. In 1976, batch production was automated through the use of flexible manufacturing systems (FMS). In 1979, wide-scale automation of design and drafting through CADD began. In 1985, the beginnings of integration or CIM began to appear. The goal of this phase is to have a completely automated manufacturing plant that operates with only a minimum of human involvement. Progress is being made in this regard, but the realization of this dream is still somewhere in the future. The wholly automated factory will combine the advantages of both the manual and mechanization phases. It will be capable of mass production, and it will be fast. It will also be completely integrated.

The most important development with regard to automation in manufacturing has been the computer. This device, coupled with other technological developments in machines and in materials handling, will eventually become the catalyst in bringing about the completely automated factory. Progress toward CIM has added a number of new terms to the language of manfacturing. For many people, these buzzwords are confusing and difficult to put into an understandable frame of reference. Doing so is necessary if one is to understand CIM.

THE BUZZWORDS AND WHAT THEY MEAN

CAD/CAM has become a frequently used buzzword in the world of design and manufacturing. So have several others including CADD, CNC, FMS, Robotics, and CIM. Before undertaking an in-depth study of CIM, one should sort these terms out and understand what they mean and how they relate to one another. (Additional buzzwords include AMH, MP & CS, CAE, APP, APPS, MRP, AI, ASR, and CAPP. These and many others have been listed in the glossary.)

What is CAD/CAM?

CAD/CAM is a broad term under which all of the other terms fall. CAD, by itself, means *computer-aided design*. Computer-aided design is a concept which encompasses any use of the computer to enhance or aid in the design process. Computers have several features that make them valuable aids to designers. These include calculation, analysis, review, modeling, and testing capabilities.

CAM, by itself, means *computer-aided manufacturing*. Computer-aided manufacturing is a concept which encompasses any use of the computer to enhance or aid in any manufacturing process. The two best-known uses of the computer to aid in manufacturing are CNC (computer numerical control) and Robotics. Almost any type of machine used in manufacturing—lathes, mills, drills, saws, punches, shears, and so on—can be computer controlled. Industrial robots are also computer controlled.

In addition to the CAM concepts of CNC and Robotics, there are other less known CAM concepts. These include computer-aided production scheduling, computer-aided quality control, computer-aided purchasing, and computer-aided sales. All of these concepts are dealt with in this book.

Using the terms CAD and CAM together (CAD/CAM) is an attempt to show the close relationship of these concepts in a manufacturing setting. It also symbolizes a goal of the automation phase of development. That goal is the elimination of the wall between design and manufacturing that was put up by specialization during the mechanization phase. When true CAD/CAM exists, there will be instant and continuous communication between design and manufacturing systems. Such communication will be electronic rather than through hard copy such as drawings.

Such communication will be accomplished over NETWORKS. Networks are the communication channels which connect design systems to other design systems, manufacturing systems to other manufacturing systems, design systems to manufacturing systems, and design/manufacturing systems to systems in other components such as sales, purchasing, shipping and receiving, contracts and bids, estimating, and accounting.

At this point in the development of automation phase, networks are the weak link in CAD/CAM. Incompatibility between CAD and CAM systems prevents the continuous communication that is necessary for the complete and instant sharing of the same database between design and manufacturing. When these problems are overcome, we will have CIM.

What is CADD?

On occasion one will see the term CADD used. This means *computer-aided design and drafting*. It is a term that evolved partially in an attempt to solve the problem of whether CAD means computer-aided design or computer-aided drafting. It also resulted from the evolution of CAD systems from the earlier models, which were strictly for enhancing the design process, to today's models, which produce design documentation in addition to aiding designers in calculations, modeling, analysis, testing, and so on.

What is CNC?

CNC means *computer numerical control*. It is one of the better-known forms of CAM. Numerical control is a concept in which machines are controlled by programs that are interpreted by a reader rather than manually, as they were during the mechanization phase. Computer numerical control (CNC) is an advanced form of numerical control in which programs may be written on and stored in a computer.

What is Robotics?

An industrial robot is a computer-controlled electromechanical device used to perform certain manufacturing tasks traditionally performed by humans. Robotics is probably the best-known form of CAM because of its controversial nature and its long-standing starring role in science fiction.

What is CIM?

CIM means *computer integrated manufacturing.* This concept represents the ultimate in manufacturing automation. CIM systems will be the key components in the wholly automated factories of the future. In fact, at some point in time, the term CIM will replace the term CAD/CAM. True CIM is the total integration of the various individual CAD/CAM concepts of CAD, CNC, Robotics, computer-aided process planning, computer-aided quality control, and materials handling. With CIM fully developed, the manufacturing continuum mentioned earlier will have gone full circle back to the total integration which existed in the days of the early craftsman.

In a fully developed CIM system, CAD systems will be networked with computer-controlled manufacturing systems which might include CNC machines, robots, and a materials-handling component. All components of the system will share the same database and have instant access to it. There will be immediate and continuous feedback at every stage of operation.

CIM is in its infancy at present. As you will see later in this book, there are CIM systems currently in place and operating. However, many are limited systems that have integrated only some of the manufacturing processes necessary to produce certain products. The total integration of all processes, from design to production to quality control to shipping of the product, is a goal that is yet to be realized.

FMS means *flexible manufacturing system* and is a term sometimes used synonymously with CIM. In reality, a flexible manufacturing system is one type of CIM system. The others are *transfer lines* and *work cells.* As you will see later in this book, transfer lines rate high in productivity and low in flexibility. Flexible manufacturing systems have medium productivity and medium flexibility ratings. Work cells have low productivity and high flexibility ratings.

The computer is the technological development that is at the heart of computer integrated manufacturing. In order to understand CIM, one must understand computers. This chapter will help students develop an understanding of computers that will, in turn, serve as the foundation upon which to build an understanding of CIM.

The functional term in the phrase "Computer-Aided Design, Drafting, and Manufacturing" is *Computer*. The computer is the technological development that has enabled the corresponding development of computer-aided design, drafting, and manufacturing, or CAD/CAM. In order to understand CAD/CAM, one must first understand computers.

This chapter presents a comprehensive overview of computer technology. It contains the information design, drafting, and manufacturing personnel need to know in order to be computer literate.

Major Topics Covered

- A Definition of Computers
- Historical Development of Computers
- Computer Hardware
- Computer Memory
- Data Representation
- Computer Software
- Computer Firmware or Microcode
- Computer Operation

Chapter One

Computer Technology

case study

One of the hot topics in manufacturing today is Computer Integrated Manufacturing (CIM). CIM, like other currently popular topics such as MRP, FMS, and JIT, means different things to different people. Some people believe that CIM is akin to science fiction and is far off in the future. Others view it as an instant fix to all manufacturing problems.

This paper will describe the work done at Cone Drive during the last five years implementing a computer system to improve our competitive position, free up working capital through inventory reduction, and improve our overall productivity.

Background

When we started this program five years ago, we had never heard of the term CIM. What we had was a series of business problems that we had to overcome in order to stay in business. The problems were the same that face many companies: too much inventory, part shortages in assembly, machine bottlenecks, missed customer delivery promises, inaccurate bills of material, and a constant barrage of Rush, Hot, Critical, and Urgent orders on the shop floor. In addition, like many of you, our marketing department was saying that we had to shorten our lead times, and our corporate people were saying that we needed to reduce inventories, "or else." To solve these problems we assembled the top management of our company. They determined that the best and perhaps only long-term solution to our current and future problems was to increase our use of computers.

Our business and our company had a major influence on our implementation methodology, although we are not at all unique. Cone Drive is a part of Ex-Cell-O Corporation, a billion-dollar corporation which primarily serves the capital goods markets. Decision-making is essentially decentralized. However, capital budgets must be justified to and approved by corporate management.

Courtesy of SME Technical Paper #MS86-724 "Building Toward Computer Integrated Manufacturing" by Paul Brauninger and John Welch, 1986.

Building Toward
Computer Integrated Manufacturing

Cone Drive markets, engineers, and manufactures double enveloping worm gearsets and speed reducers to a variety of both domestic and international customers. We sell both standard and special units, but only assemble a finished reducer after receipt of a customer order. We are basically a job shop employing about 350 people, with sales between 25 and 50 million dollars annually.

Implementation

Our plan for computerization had four basic criteria: acquire canned programs, purchase hardware and software from a single source, train all employees thoroughly, and change our business practices to the software requirements whenever practical. In addition, we decided on a plan that required us to build a good foundation of data and reporting systems before entering areas of new technology. Because of increased customer dissatisfaction with lead times and deliveries, we first started looking for a good Inventory/Production Control system.

As soon as we started looking into MRP systems, it became clear that we needed a new integrated business system. This needed to cover all areas of our business in order to give us the maximum benefit. We selected a software package called MAPICS that ran on an IBM System 38 computer.

The MAPICS package included all of our business functions: MRP, routings, bills of material, job costing, inventory management, order entry, invoicing, accounts receivable, accounts payable, payroll, capacity planning, and shop floor data collection. We installed this package in January 1982, after planning and training for eighteen months. This planning included the rental of an IBM System 38 computer to train employees and test out the software. Because of the planning that had gone on before—and, of course, a lot of luck—we were able to install all the modules at one time without serious disruption of the manufacturing floor

(continued)

or office. This system gave us the fundamental building blocks upon which to build other systems.

As time passed, we found that our business system allowed us to reduce inventories, cut our end product lead times, and significantly improve our delivery integrity. But we also found that in some instances it was taking us longer to release an engineering drawing than it took to manufacture the part. Because of this, we decided to install a CAD/CAM system (CADAM).

We made a basic decision that the MAPICS bill of material and part number file would be the master and the CADAM system would handle the drawing and N.C. programming area only. After testing and evaluation of several systems, we installed our CADAM system in January 1984 and found it very beneficial, not only in reducing the drawing time an average of 60% but also in allowing us to reduce our engineering lead time by working multiple shifts. The idea of working multiple shifts came about because of the high fixed costs of the system that was very similar to an N.C. machine tool. We decided that we would use the same approach to maximize the output of our CADAM system that we used for many years in manufacturing.

Shop Floor Integration

In late 1983, when CADAM was being tested, we found that we had accomplished and were accomplishing a great deal by computerization of our business and engineering systems; but we needed to further improve productivity on our shop floor. We decided to attempt to integrate our two systems, which had been run on two separate computers, to obtain more accurate, timely, and complete information. We accomplished this by putting Personal Computers on the shop floor. These are used for DNC (replacing paper tape), engineering drawing inquiry (viewing drawings on PC), and message and scheduling communication. The use of the PC has proven to be very beneficial from a cost and productivity viewpoint, and as an employee morale factor. Now we can provide the individual machine operator with up-to-date information on what to run, visual data on machine set-up, engineering drawing data, and other manufacturing, machining, or inspection data.

We find that we are constantly uncovering ways to improve productivity through the use of the computer. MRP, JIT, and FMS are all tools to achieve a more efficient organization. The key is in planning and using the available tools to their maximum. With the advances in computer hardware, technology, and the relatively low cost of canned software, our challenges will be how to integrate these new and old technologies into our particular business to ensure the long-term viability of that entity.

Summary

CIM is not a magic wand that can be waved over a company to produce solutions to every problem. Nor is it a solution that will happen without effort. CIM is made up of many parts that take planning, organization, and control to be successful. But, if you are willing to make that commitment, the results are impressive. Those companies that are going to be successful in the future must start building towards CIM today. As we all know, we compete in a global economy, and, like the dinosaur, those companies that do not adapt to changing times will cease to exist.

A DEFINITION OF COMPUTERS

A computer is an electronic machine that has storage, logic, and mathematical capabilities and is able to perform certain tasks at extremely high speeds. A computer has four characteristics which distinguish it from other machines.

1. Computers perform all operations electronically.
2. Computers have an internal storage capability.
3. Computers receive operational instructions from stored programs.
4. Computers can modify program executions by making logical decisions.

In analyzing these four characteristics, it is important to remember that such devices as disk drives, keyboards, and other peripherals are not themselves computers. These devices are electromechanical devices and would, on the surface, seem to violate the first characteristic of computers. However, one must remember that these are peripheral devices which allow humans to interact with computers but are not in themselves the computer.

The human/computer team is a powerful one in terms of productivity in getting work done. The computer has two capabilities that make it particularly valuable as a tool for human use: A computer is extraordinarily fast as compared to human beings, and a computer is much more accurate and reliable than a human being.

However, the computer has two critical shortcomings when compared with humans. First, a computer cannot reason and think as humans can. Computers are capable of making decisions based on mathematical logic, but they cannot apply common sense, make judgments, or use intuition. Second, a computer cannot adapt or innovate during the problem-solving process. A computer that has been incorrectly programmed will simply continue on its charted course, making the same mistakes, regardless of circumstances, until it is shut down or reprogrammed.

Classifications of Computers

The two main classifications of computers are *digital* and *analog*. The type of computer most familiar to people and most frequently used is the digital computer. Design, drafting, and manufacturing applications use digital computers.

A *digital computer* receives input in the form of numbers, letters, and special characters, such as those found on a typewriter keyboard. It converts these numbers, letters, and special characters into electronic binary signals for processing. During the output stage of operation, the binary signals are converted back into numbers, letters, and special characters that humans can understand. Graphic data, such as that used in design and drafting, are also converted into binary form as they are entered into a digital computer, and then converted back to graphic form as they are output by a digital computer.

Analog computers are not used in human interaction settings, and they do not operate on binary digits. Rather, they equate electrical analogs and such physical magnitudes as heat, pressure, and fluid flow. It is not important that design, drafting, and manufacturing students understand what an analog computer is. Rather, it is important that they understand that analog computers are not the type used in design, drafting, and manufacturing, and that digital computers are.

The computers used in modern design, drafting, and manufacturing settings are much more sophisticated and much more powerful than their earlier predecessors. In order to be computer literate, CAD/CAM students should understand the historical development of computers.

HISTORICAL DEVELOPMENT OF COMPUTERS

Prior to the invention of the computer, those tasks now performed by computers had to be performed either manually, which means by hand, or mechanically, which means by machine. An easy way to understand the significance of the development from manual to mechanical to computer processes is to consider the example of solving math problems. There was a time when math problems had to be solved using longhand. Problems were set up and solved using a pencil and paper.

This involved such processes as addition, subtraction, multiplication, and long division. Solving mathematical problems manually was a long, involved, and arduous task. If you don't remember just how difficult a task this was, refresh your memory by doing a long division problem (such as 9,472.16 divided by 348.52) manually. In the next generation of development, mathematical problems could be solved mechanically with the assistance of slide rules and adding machines.

Mechanical processes represented a major improvement over manual processes. However, they still left much to be desired. Finally, the electronic calculator emerged; with it, the solution of mathematical problems, regardless of their complexity, became a simple task. Raising numbers to powers, taking square and even cube roots, and dividing even the longest numbers were no longer difficult tasks.

This same type of development has occurred in the fields of design, drafting, and manufacturing as processes have moved from the manual to the mechanical and, finally, to the computer stage. The computer, as defined earlier, had its birth in the early 1940s. One of the first computer prototypes was developed by a team of Harvard University graduate students working in conjunction with IBM engineers.

This first prototype of the computer was known as Mark I. By today's standards, it would be considered a dinosaur; big and slow. The Mark I electromechanical computer was over fifty feet long and eight feet high. It weighed over five tons. The Mark I required approximately six seconds to multiply two numbers having up to twenty-three digits each. It could divide the same numbers in just a little over five seconds. Of course, by today's standards, these times are very slow. In Mark I's time, however, they were thought to be revolutionary.

At the same time that IBM was working with Harvard University developing the Mark I, a professor at Iowa State College (now Iowa State University) was working with one of his graduate students to design and build a computer. Their prototype was completed in 1942 and came to be known as the ABC or Atanasoff/Berry Computer, taking its names from the last names of its inventors, Professor John Atanasoff and graduate assistant Clifford Berry.

The next major step in the development of computers was ENIAC. ENIAC was the acronym for Electronic Numerical Integrator and Calculator. ENIAC used vacuum tubes instead of electromagnetic relays for switching and control functions. This increased its speed over earlier prototypes significantly. ENIAC was still a monstrous machine by today's standards. It weighed thirty tons and took up approximately 1500 square feet. In spite of this, ENIAC got considerably more use than its predecessors. It was used in ballistic research by the U. S. Army from 1946 to 1955. Today, it can be seen preserved for posterity in the Smithsonian Institution in Washington D.C.

The first computer to be used commercially was developed by the Sperry Rand Corporation and purchased by the U. S. Government for use in the Bureau of the Census in 1951. This computer was called UNIVAC I. The first private-sector or business use of a computer occurred in 1954, when a branch of the General Electric Corporation purchased its first computer. The historical development of computers from ENIAC to the modern computers of today is divided into generations.

Generations in the Development of Computers

The development of computers has been inseparably tied to developments in the field of electronics. The state-of-the-art electronic device when computers such as UNIVAC were developed was the *vacuum tube*. Consequently, the vacuum was the functional enabling device in the first generation of computers. Computers containing vacuum tubes were large cumbersome devices, as we have already seen. Those of us who are old enough can remember televisions and radios which used vacuum tubes, and we can compare their sizes with those of the models available today.

The next critical development in the evolution of electronics and, correspondingly, the evolution of computers was *solid-state circuitry*, based on the *transistor*. Transistors were considerably smaller than vacuum tubes. Consequently, technological developments relying on transistors—such as the television, the radio, and more particularly, the computer—also became smaller; they also became more powerful. Computers using the transistor as the functional enabling device were considered second-generation computers.

The significant technological development in electronics and computers that followed the transistor was the *integrated circuit*. Scientists and engineers working for various electronic manufacturers discovered that over six hundred transistors, diodes, and other electronic components could be placed on one single slice or wafer of silicon. Integrated circuits on silicon wafers, sometimes called "chips," were much smaller than transistors.

Technological developments which had previously relied on the transistors as the functional enabling device now switched to integrated circuits. Once again, the pattern of smaller yet more powerful computers resulted. Computers which rely on integrated circuits are known as third-generation computers. You will recall that the ENIAC computer weighed thirty tons and took up 1500 square feet of space. Today, an even more powerful computer could be made by mounting integrated circuits on just one printed circuit board to produce a computer no larger than a portable television set, such as the one shown in Figure 1-1.

We are still in the third generation of computers. However, integrated-circuit technology has continued to improve. The first integrated circuits were limited in that they could only be used for the specific purpose for which they were designed. In the early 1970s, however, an engineer working for the Intel Corporation developed a more advanced integrated circuit that was programmable and could be used for a number of purposes.

This programmable integrated circuit was a tiny processing unit unto itself. Hence, it came to be known as a *microprocessor*. It was the microprocessor that led to the computer revolution as we know it today. Microprocessors are now used in microwave ovens, wristwatches,

FIGURE 1-1 The type of printed circuit board found in today's small but powerful computers.
Courtesy of Vermont Microsystems, Inc.

telephones, televisions, radios, automobiles, aircraft, spacecraft, and hundreds of other diverse applications.

Today, microprocessor design and production is a big business by itself. There is intense competition among a number of firms such as Bell Laboratories and Hewlett-Packard to continually produce better yet less expensive microprocessors. The goal of scientists working in microprocessor development is to produce a single-chip microprocessor that is packed with electronic components as densely as the human brain is packed with neurons.

COMPUTER HARDWARE

Machines and equipment in a computer system are called hardware. Hardware can be divided into four broad categories: 1) the processing unit, 2) secondary storage devices, 3) input devices, and 4) output devices (Figure 1-2). A fifth category, input/output devices, can be added. Input/output devices, naturally, are devices that can perform both input and output tasks. The processor, secondary storage devices, input devices, and output devices all have separate functions to perform in a computer system. However, each relates to the others in certain specific and important ways.

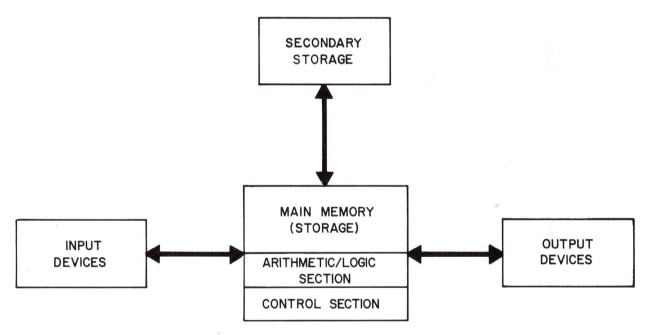

FIGURE 1-2 Computer hardware.

The Processing Unit

The processing unit in the computer consists of two major components: the control section and the arithmetic/logic section. It should be noted that while the main memory is often housed in the same console as the two components of the processing unit, it is not contained on the same printed circuit board (except in the case of the microcomputer). Consequently, the control section and the arithmetic/logic section taken together are called the central processing unit of the computer. When the control section and arithmetic/logic section are both placed on one single silicon chip, you have a microprocessor.

The purpose of the control section is to direct all activities of the computer as set forth in stored programs. It controls the arithmetic/logic section, input and output operations, the transmittal of data to and from storage, and all other operational activities of the computer. The arithmetic/logic section performs all addition, subtraction, multiplication, division, and other mathematical operations; it also compares data so that logical decisions such as "if-then" decisions can be made by the computer.

You will recall that one of the four characteristics that sets a computer apart from other machines is that it can modify program instructions by making logical decisions. A computer has this capability because the arithmetic/logic section of the central processing unit is able to effect the comparison needed for making such logical decisions.

The main memory is the primary repository for the storage of computer programs and data being processed by the computer. At any given time, the main memory of a computer might contain data that are awaiting processing, data that are being processed, data that have been processed and are waiting to be output, and stored programs that guide the operation of the computer.

Secondary Storage Devices

No matter how large the main memory of a computer, it is never large enough to contain all of the data with which it must deal. Additionally, main memory is one of the most expensive aspects of a computer system. An alternative to simply expanding the size of the main memory of the computer system is to use secondary storage devices. Secondary storage devices, sometimes called auxiliary storage devices, are such devices as magnetic tape units, magnetic disk units, and floppy disk drives (Figure 1-3).

Secondary storage devices offer relief to the main memory. Programs and data that have been processed can be saved on a secondary storage medium and stored *off-line*. Data that are stored in the

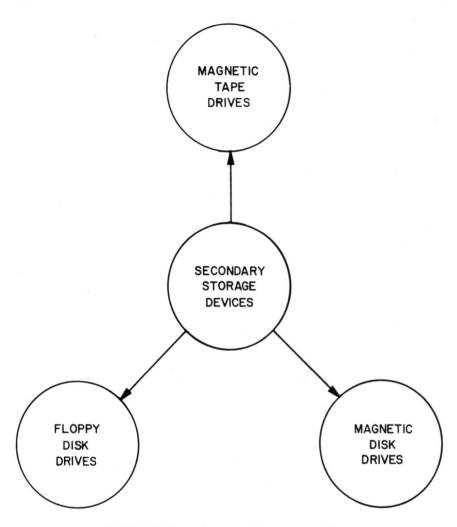

FIGURE 1-3 Secondary or auxiliary storage devices.

main memory are said to be *on-line*. On-line data are immediately available. Data stored off-line are not immediately available.

Input and Output Devices

Input and output devices are devices which allow humans to interact with the computer in such a way as to enter into the computer data that are to be processed and to take out of the computer data that have been processed. Input devices are especially designed to convert ingoing data into computer-usable form. Output devices are especially designed to convert data coming out into human-usable form.

There are a number of different types of input and output devices. Some of the most frequently used are keyboards, display terminals, magnetic tape units, teleprinters, line printers, optical bar code readers, optical mark readers, optical character recognition readers, and magnetic ink character recognition readers.

Computers use keyboards that are very much like the keyboards on typewriters, with keys for letters, numbers, and a variety of special characters. Keyboards are used for inputting alphanumeric data, programs, and other special symbols interactively. The keyboard is probably the most widely recognized and frequently used input device.

Display terminals are output devices for displaying alphanumeric and graphic data in soft form. Soft form means that when the display terminal is turned off, the data go away. This is in contrast to hard copy, which is any type of output that can be taken away from the computer, such as a printout. A display terminal looks a lot like a small television. Many computer models combine a display terminal and a keyboard in the same console unit.

Magnetic tape drive units, or magnetic tape drives, are input/output devices. Magnetic tape is a form of auxiliary or secondary storage. Magnetic tape can be used for storing programs, as well as data that have been processed. The magnetic tape drive allows data that have been stored on magnetic tape to be entered into the computer at high speeds, and it allows data that have been processed to be removed from the computer and stored off-line.

A line printer is a typewriter-like device that produces alphanumeric hard copy. The product produced by a line printer has come to be known as a "printout." Line printers may be of the *impact* or *dot matrix* varieties. An impact line printer creates alphanumeric symbols in the same way that a typewriter does; keys strike a ribbon, forcing an image onto the paper. Dot matrix printers create alphanumeric characters by arranging tiny dots in various patterns within an established matrix.

Most modern line printers are bi-directional, meaning that they print one line moving from left to right and print the following line moving back from right to left. This, of course, cuts the printing time in half. A high-speed line printer can print over 1000 lines per minute. Some of the more advanced, more expensive line printers can print as many as 10,000 lines per minute. Teleprinters can be used for both input and output.

Magnetic ink character recognition devices are used in such applications as banking or reading large numbers of checks in a short amount of time. Such readers interpret the individual magnetic ink characters on checks or any other type of document.

Optical character recognition readers read the patterns established by individual characters, such as letters and numbers, and convert these into electrical impulses. Another optical reading device is the optical mark reader. Optical mark readers sense positions of marks on a document, rather than sensing the form or shapes of characters. Such readers are popular in the scoring of standardized tests. Yet another optical reading device is the optical bar code reader, which senses the shape, length, and width of a variety of shaded bars, such as those found on packages in grocery stores. In fact, a popular use of optical bar code readers is in checkout lines in grocery stores.

COMPUTER MEMORY

The memory section of the processor is important enough to warrant special attention. There are certain features about the memory section with which CAD/CAM technicians should be familiar. These include such concepts as "bit," "byte," and "word"; overall memory size; and several miscellaneous features, such as parity, cache, and virtual memory.

Memory Location Sizes

The main memory in a computer is classified as an 8-, 16-, 32-, 48-, or 64-bit memory. These classifications refer to the size of the computer word the memory can handle. An 8-bit word size is, of course, smaller than a 32-bit word size. A word is the amount of data a computer can retrieve from memory per instruction (Figure 1-4).

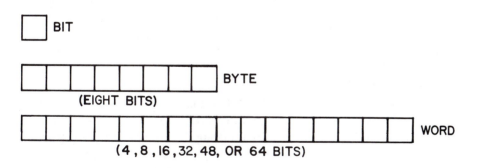

FIGURE 1-4 Comparison of relative sizes of bit, byte, and word.

A bit is a binary digit represented in the computer by electrical impulses that are either *on* or *off*. On paper, binary digits are represented by the numerals "1" and "0." A one means "on." A zero means "off."

If each bit in a 32-bit memory (a common size in CAD system processors) is considered to be a switch that can be turned on or off (two possibilities), there are over 2,000,000,000 possible combinations of on-off sequences. This means that a 32-bit memory location can handle a word of 2,147,483,648 bits, which is 2 raised to the thirty-first power. It is not 2 to the thirty-second power because one bit is always saved and not used.

Overall Memory Size

The overall size of a CAD system's memory section depends on the number of memory locations it has. Because the numbers can be so large, memory sizes are stated in Ks. The letter K is short for "kilo," which means 1,000. However, the actual multiplier used in sizing computer memories is 1,024. A 64K-byte computer has 64 X 1,024 or 65,536 bytes of memory. Large computers use the prefix mega (1,000,000) or megabytes.

Depending on the manufacturer of the computer, a byte can be 6, 7, or 8 bits. However, it is common practice to consider that one byte equals 8 bits (Figure 1-4). The term RAM, when used in conjunction with computer memory, means "Random Access Memory" (i.e., 64K bytes of RAM). This is "read-write" memory, in which the computer can go directly to the desired data, regardless of its storage location.

Miscellaneous Memory Features

Overall memory size and word size are the two most important features to understand about the memory section of a CAD system's processor. However, there are several other memory features with which CAD/CAM technicians should be familiar. These include mapping, parity, and cache and virtual memory.

Memory Mapping

Memory mapping is a function that is used to guide programs to data in selected sectors of the main memory while keeping them out of other sectors of memory. Memory mapping serves the same purpose as giving a person specific instructions regarding which mailboxes to empty at a post office: It cuts down on the time it takes to locate and retrieve the desired mail, and it protects mail in the other boxes.

In Figure 1-5, the memory map would allow a program to access data in storage locations 4, 9, 12, 13, 14, 16, 17, 18, and 19, while protecting the data in all other memory locations. This, in effect, increases the amount of time required to access the data, but it decreases the overall operating time by providing the processor with exactly the data required.

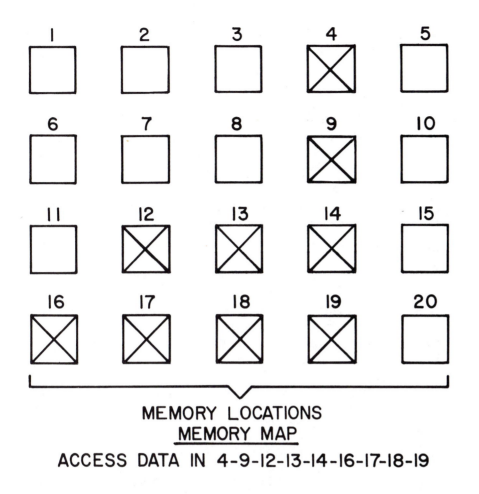

MEMORY LOCATIONS
MEMORY MAP
ACCESS DATA IN 4-9-12-13-14-16-17-18-19

FIGURE 1-5 The concept of memory mapping.

Parity

A line taken from a manufacturer's hardware specifications reads: "Main memory—256 kilobytes RAM with parity." Translated, we know that this means 256K bytes of random access memory with parity. *Parity* is an extra bit in each byte that is used for detecting errors in the data contained in each memory location.

It works as follows: Data within individual memory locations are represented by a series of "on" and "off" signals. As we learned earlier, "on" signals are represented on paper by "1" and "off" signals by "0." Parity bit error detecting uses either the odd-parity or the even-parity method to check data.

In an even-parity check, the computer knows that the sum of all 1s or "ON" signals in an individual memory location plus the parity bit will be an even number. In an odd-parity check, the sum of all 1s or "ON" signals plus the parity bit will be an odd number. When these rules are not satisfied, the computer knows that an error exists.

Cache Memory

Cache memory is a small, fast, additional memory used to increase the speed of interaction and the processing speed of a computer. Cache memory is a simple concept. Frequently accessed data are taken from the main memory and stored in the cache memory. Because the cache memory is smaller than the main memory and contains only selected data, the processor can access data contained there in less time and with less searching. This, of course, speeds processing time.

Virtual Memory

Virtual memory is a concept used when extremely long programs would take up more memory space than is desirable. When dealing with long programs, virtual memory allows a sector of the main memory to be designated as the sector to accept the program. The program is then broken into sections that will fit within the designated sector of memory. The remainder of the program is stored on a secondary storage device, usually a disk.

Each time a successive piece of the program has been used and the next section is required, the new section is transferred from the disk and overlaid on the designated sector of main memory. The new section actually replaces the old section in memory. This process is repeated until the entire program has been run. Virtual memory causes the main memory to appear to have a larger capacity than it actually has.

The preceding paragraphs have talked about computer memory in terms of size and the various ways it is arranged. In the computer hardware section, you learned about the main memory of a computer system and secondary storage devices. There are a number of different types or classifications of computer memory. The simplest way to classify the types of computer memory is as being main memory and auxiliary or secondary memory. In this way, the types of computer memories correspond with the hardware about which you have already learned.

Main Memory

The main memory of a computer is actually directly connected to the central processing unit and usually housed in the same console as the central processing unit. Main memory can be divided into three categories:

1. Main Data Storage
2. Control Storage
3. Local Storage

Main data storage is what makes the main memory of a computer expensive. There are a variety of technologies used for main data storage. One frequently used technology is *magnetic core storage*. Another popular main data storage technology is *semiconductor storage*. In semiconductor-storage technology, memory locations are made up of transistor circuits.

Control storage assists the control section in carrying out its tasks. Local storage is used by the arithmetic/logic section.

A final main memory technology is *semiconductor monolithic storage*, sometimes called large-scale integrated memory circuitry, or LSI memory circuitry. This type of main memory consists of thousands of miniaturized transistor circuits.

Auxiliary or Secondary Storage

Main memory is expensive. The more main memory a computer has, the more expensive the computer. Consequently, in order to hold down the cost of a computer, the size of its main memory must be limited. Consequently, it is common practice to store programs and various types of data files on secondary storage devices. In this way, data can be loaded into the main memory as needed and stored off-line when not needed. Secondary storage is important because many computer programs are so long and complex that they would take up an inordinate amount of room if placed in the main memory.

You have already read that the principal secondary storage devices are magnetic tape drives, magnetic disk drives, and floppy disk drives. These are auxiliary storage devices. In addition, auxiliary storage itself can be divided into two broad categories.

1. *Sequential access auxiliary storage* is the least effective type of secondary storage. It is called sequential storage because in order to access a file in the middle of the storage device, the computer must read all files prior to it. This is like having to listen to all of the songs on a cassette tape in order to hear the last one on the tape.
2. *Direct access auxiliary storage* is much more effective than sequential access storage. As the name implies, with direct access auxiliary

storage, the computer is able to go directly to or access exactly the file it wants, without having to read any files which may precede it.

There are several widely used devices for secondary storage. Probably the one most people are familiar with is the floppy diskette, frequently used with microcomputers. Floppy diskettes are a form of magnetic disk storage. They come in a variety of sizes. However, the most frequently used size is the 5¼-inch diameter floppy diskette. Floppy diskettes, which are direct access auxiliary storage devices, are actually magnetically coded plastic disks.

Another frequently used auxiliary storage device is the magnetic tape. Magnetic tape is actually magnetically coded mylar tape that runs on a reel-to-reel basis. Magnetic tape is a form of sequential access auxiliary storage.

Another commonly used auxiliary storage device is the magnetic drum. A magnetic drum is a magnetically coded cylinder which rotates at a constant speed under a read/write head. Data are stored on the drum in the form of magnetized spots. The magnetic drum is a form of random access auxiliary storage.

At present, these are the three most commonly used auxiliary storage mediums. Other less-used and emerging auxiliary storage technologies include bubble memory, laser beam storage, video disk memory, and electron beam addressable memory.

Bubble memory stores data as tiny magnetic bubbles on extremely thin magnetic film. Bubble memory is capable of high storage and random access. However, it is still a technology that has yet to be developed completely.

Laser beam storage stores data as microscopic bits on special metal film strips. Data are read by a laser beam which scans the surface of the metal strips and measures the amount of light that is reflected.

Video disk memory systems represent a technology that is already used in television and other types of videotechnology. Video disks store data on small platters which resemble phonograph records. Video disks have an enormous storage capacity as compared to other storage technologies.

Electron beam addressable memory systems combine large-scale integrated-circuit memory chips and a cathode ray tube. Data are written to each chip and read from each chip by an electron beam. Electron beam addressable memory has a high access-time rate and a high data-transfer rate.

DATA REPRESENTATION

You may have learned in a math class that the numerals that we use (1, 2, 3, 4, 5, and so on) are not numbers but symbols which represent numbers. The same could be said of the alphabet. The alphabetical characters we use to represent specific sounds are not actually the sounds themselves but symbols which represent the sounds. In order to read and understand words which are composed of letters of the alphabet, you must be able to interpret what the symbols represent.

This is also true of computers. In order to accept, store, process, and output data, computers must have some means of data representation that they understand. There are several different methods of representing data that computers can understand. There is the binary system, the octal system, and the hexadecimal system. The most common of these is the binary system. Human users of computers seldom deal with binary, octal, or hexadecimal data.

It is not necessary to understand these systems in order to be a computer user. However, if one does understand at least one of these systems, it will improve his or her overall understanding of computers. Consequently, the following sections contain a treatment of the binary system.

Decimal and Binary Numbers

The base or "radix" of the decimal numbering system is 10. The radix of the binary numbering system is 2. All decimal numbers are represented by one or a combination of ten numbers, 0 through 9. All binary numbers are represented by one or a combination of two numbers, 0 and 1. The chart in Figure 1-6 shows the binary equivalent for decimal numbers 1 through 20.

In the decimal numbering system, each digit carries a certain "weight," depending on its position. For example, in Figure 1-7, the 3 in the "tens" position is actually equal to 3 X 10, or 30. The 3 in the "hundreds" position is actually equal to 3 X 100, or 300. Therefore, in the decimal numbering system, the value of a particular digit is found by multiplying that digit times the weight of its position.

Digit positions carry weight in any numbering system. Regardless of the system, the weight of a digit position can be found by multiplying the weight of the digit position immediately to the right by the radix of the numbering system (10 for the decimal numbering system and 2 for the binary numbering system). Figure 1-8 contains a weight chart for the binary numbering system. Notice from the charts that whereas

DECIMAL	BINARY	DECIMAL	BINARY
1	1	11	1011
2	10	12	1100
3	11	13	1101
4	100	14	1110
5	101	15	1111
6	110	16	10000
7	111	17	10001
8	1000	18	10010
9	1001	19	10011
10	1010	20	10100

FIGURE 1-6 Binary equivalents of selected decimal numbers.

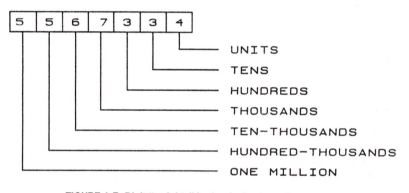

FIGURE 1-7 Digit "weights" in the decimal number system.

the positions increase in multiples of 10 in the decimal system, they increase in multiples of 2 in the binary system.

The Binary Numbering System

When working with binary and decimal numbers simultaneously, the radix is placed after the number in order to avoid confusion. For example, the decimal number 1011 can be written (1011)10 to avoid mistaking it as a binary number.

Students might want to learn how to convert back and forth between binary and decimal numbers. In either case, it is a matter of several simple steps.

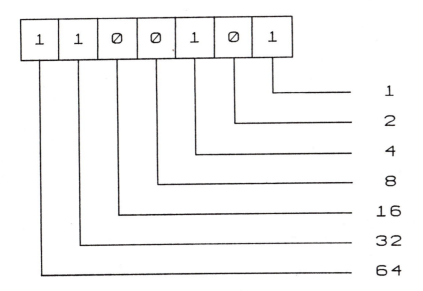

FIGURE 1-8 Digit "weights" in the binary number system.

Converting Decimal Numbers to Binary Numbers

Step 1: Begin by drawing two lines at right angles to form a large "T." This T will serve as the format for the table which will be created in the conversion process (Figure 1-9).

Step 2: Enter the decimal number that is to be converted to binary on the left-hand side of the table, as shown in Figure 1-9.

Step 3: Divide the decimal number by 2. Place the answer on the left-hand side of the table immediately under the original number, and place the remainder on the right-hand side of the table as shown in Figure 1-9. If the division works out evenly and there is no remainder, a zero is placed in the right-hand column.

Step 4: Step 3 is repeated until the last number in the left-hand column is zero, as shown in Figure 1-9.

Step 5: The binary number is read from the right-hand side of the table, starting at the bottom of the column of numbers and working up to the top, as shown in Figure 1-9.

Converting Binary Numbers to Decimal Numbers

Step 1: Invert the binary number and write it in column form, as shown in Figure 1-10.

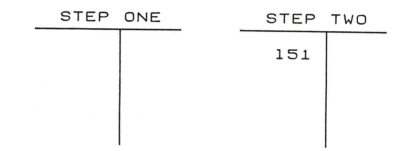

STEP ONE	STEP TWO
	151

STEP THREE		STEP FOUR	
151	1	151	1
75		75	1
		37	1
		18	0
		9	1
		4	0
		2	0
		1	1
		0	

STEP FIVE

SOLUTION = 10010111

FIGURE 1-9 Table method of converting decimal numbers to binary numbers.

Step 2: Beginning at the top of the column, place a multiplication sign to the right of each digit, as shown in Figure 1-10.

Step 3: Beginning again at the top of the column, place the appropriate weight value to the right of each digit, as shown in Figure 1-10.

Step 4: Multiply each digit by its corresponding weight value and record the answer, as shown in Figure 1-10.

Step 5: Add the new column of multiplication products together and read the decimal number, as shown in Figure 1-10.

```
BINARY NUMBER = 10010111
```

$$1 \quad \times \quad 1 \quad = \quad 1$$

$$1 \quad \times \quad 2 \quad = \quad 2$$

$$1 \quad \times \quad 4 \quad = \quad 4$$

$$0 \quad \times \quad 8 \quad = \quad 0$$

$$1 \quad \times \quad 16 \quad = \quad 16$$

$$0 \quad \times \quad 32 \quad = \quad 0$$

$$0 \quad \times \quad 64 \quad = \quad 0$$

$$1 \quad \times \quad 128 \quad = \quad \underline{128}$$

$$151$$

FIGURE 1-10 Converting binary numbers to decimal numbers.

The Octal Numbering System

The octal system is a base 8 system which uses the following characters:

0, 1, 2, 3, 4, 5, 6, and 7

The octal number system is a convenience system. It is used as an alternative to the binary system in which numbers can become unreasonably long. Compare the following binary numbers and their octal equivalents:

Octal Number	Binary Number
5	101
10	1000
15	1101
20	10000

Each character in an octal number can be represented by a three-digit binary number. Such a grouping can be used for converting binary numbers to octal numbers.

Converting Binary Numbers to Octal Numbers

Binary numbers can be easily translated into octal numbers by arranging them into groups of three, beginning with the least significant digit and replacing each group of binary digits with its octal equivalent. The octal characters and their binary equivalents are shown below.

Octal Characters	Binary Grouping
0	0
1	1
2	10
3	11
4	100
5	101
6	110
7	111

Using this method, the binary number (1110101111101) would be translated as follows:

Step 1: Beginning with the rightmost or least significant digit, arrange the binary number into groups of three digits.

<u>001</u>   <u>110</u>   <u>101</u>   <u>111</u>   <u>101</u>

Step 2: Replace each group with its octal equivalent.

$$001 \quad 110 \quad 101 \quad 111 \quad 101$$
$$(1) \quad (6) \quad (5) \quad (7) \quad (5)$$

Step 3: The octal equivalent of 1110101111101 is

16575

Notice in Step 1 that it was necessary to add two insignificant zeros to the left of the one in order to have a group of three. The reverse of this process can be used to convert octal numbers to binary numbers, as shown below.

Step 1: Arrange the octal number with sufficient space under each character to allow for placement of corresponding binary numbers.

$$1 \quad 6 \quad 5 \quad 7 \quad 5$$

Step 2: Arrange the binary equivalent of each number under its corresponding character.

$$1 \quad 6 \quad 5 \quad 7 \quad 5$$
$$(001) \quad (110) \quad (101) \quad (111) \quad (101)$$

Step 3: The binary equivalent of 16575 is

1110101111101

The Hexadecimal Number System

The hexadecimal system is an extension of the octal system in that it can be used to simplify binary numbers even further. The hexadecimal system is a base 16 system which uses the following characters:

0, 1, 2, 3, 4, 5, 6, 7, 8, 9, A, B, C, D, E, and F

One hexadecimal character can represent four binary digits. This means a long binary number can be represented by a relatively short hexadecimal number. For example, the binary number 10101011111101 can be represented by the hexadecimal number 2AFD. The binary equivalents of the hexadecimal numbers are shown below.

Hexadecimal Number	Binary Number
0	0
1	1
2	10
3	11
4	100
5	101
6	110
7	111
8	1000
9	1001
A	1010
B	1011
C	1100
D	1101
E	1110
F	1111

Using this list of equivalents, it is a simple process to convert long binary numbers to relatively short hexadecimal numbers. The binary number 10101011111101 is converted to hexadecimal as follows:

Step 1: Beginning with the least significant digit, arrange the binary number into groups of four digits.

0010   1010   1111   1101

Step 2: Place the hexadecimal equivalent for each group under it.

0010   1010   1111   1101
(2)   (A)   (F)   (D)

Step 3: The hexadecimal equivalent of the binary number is

2AFD

Of course, the process can be reversed for converting hexadecimal numbers to their binary equivalents.

Binary Coding Schemes

You have just seen how numbers can be represented in a computer using the binary numbering system. However, numbers are not the only data with which computers must deal. There are also letters and special characters such as those found on a typical keyboard.

There are several binary coding schemes which allow computers to deal with numbers, letters, and special characters. The three most common binary coding schemes are the American Standard Code for Information Interchange (ASCII), the Extended Binary Coded Decimal Interchange Code (EBCDIC), and the Binary Coded Decimal System (BCD).

ASCII Coding Schemes

The ASCII coding scheme is used primarily in telecommunications applications. It is a 7-bit coding scheme which allows for 128 different bit patterns for representing various types of characters. It was developed to simplify machine-to-machine and system-to-system communication.

EBCDIC Coding Scheme

The EBCDIC coding scheme is capable of representing 256 unique and distinct characters. These characters can be numbers, letters, and special characters.

BCD Coding Schemes

The BCD coding scheme can represent a maximum of 2 to the sixth power, or 64 distinct characters. The BCD coding scheme uses a total of 7 bits. The first 6 bits represent data that can be numbers, letters, or special characters. The final bit position is a parity check bit.

COMPUTER SOFTWARE

Software is a generic term used to describe the nonmechanical components of a computer system. These components include computer programs, documentation of those programs, and the various types of technical and reference manuals that go with the system. However, when most people use the term "software," they are talking strictly about computer programs. A computer program is a specially coded set of instructions that directs the computer in performing all operations.

People who are trained to write programs are called "computer programmers." It is not necessary to know how to program a computer in order to be an efficient and effective computer user. Most computer users do not know how to program a computer. However, it is helpful if one understands what computer programs do and something about the various languages in which computer programs are written. The next section presents some of the more commonly used computer programming languages.

Computer Programming Languages

Computers are only capable of doing what they are programmed to do. People write computer programs. Consequently, people control computers. Communicating with a computer via a computer program is similar to talking with a person from another country. In order to communicate, one must use that person's language, or a language he or she understands.

There are over 300 computer programming languages that various types and models of computers can understand. The reason there are so many different computer programming languages is that there are so many different applications and uses of computers. There are three levels of computer programming languages. The first level is represented by machine language, the second consists of assembly languages, and the third consists of a variety of high-level programming languages.

Machine Language

Machine language is the actual language of a computer. In human terms, machine language is at the lowest level possible. It actually consists of the presence and absence of the electrical impulses upon which a digital computer operates. Machine language is not actually just one language. Rather, the language can vary for each type and model of computer. The important thing to remember about machine language is that it can be understood directly by the internal circuitry of the computer. It requires no interpretation.

Assembly Languages

The assembly languages are still considered low-level languages. Assembly languages are written using special symbols or *Mnemonics*. Mnemonics are memory aids which make it simpler for humans to write, enter, and test computer programs. A program written in an assembly language must be converted into machine language before it can be used by a computer.

This is accomplished by an assembler program, which is a program that assembles the various program statements written in assembly language and converts them to a machine language format. This is accomplished by substituting absolute operation codes for the symbolic operation codes of the assembly language and absolute or relocatable addresses for the symbolic addresses contained in the assembly language.

When an assembler program converts a program that has been written in an assembly language into machine language, it also produces an assembly listing. An assembly listing is a printed copy of an assembly run which shows the source program, or the assembly language program, and the object program, or the machine program.

High-Level Languages

Assembly languages are machine-dependent. In other words, the programmer must know not only the assembly language but also a great deal about the actual computer with which the program will be used. High-level computer programming languages, on the other hand, are procedure-oriented rather than machine-oriented. They are relatively independent of the computer upon which they will be run. There are numerous high-level computer programming languages. Some of the most commonly used of these are BASIC, COBOL, FORTRAN, APL, RPG, PL/1, and PASCAL (Figure 1-11).

BASIC stands for Beginners All Purpose Symbolic Instructional Code. It is a high-level programming language designed primarily for use in interactive problem-solving situations. It is a simple language, easy to learn, and easy to understand. Therefore, it is one of the more frequently used languages.

COBOL stands for Common Business Oriented Language. COBOL is a high-level programming language designed primarily for business data-processing applications.

FORTRAN stands for Formula Translator. FORTRAN is a high-level computer programming language designed primarily for scientific and engineering applications in which a high number of mathematical computations will be necessary.

APL is the acronym for A Programming Language. APL was designed primarily for use in interactive problem-solving situations.

RPG is the acronym for Report Program Generator, which is a high-level programming language designed for applications that involve a large number of printed reports and other similar types of documents. Like BASIC, RPG is an easy computer programming language to learn.

PL/1 is the acronym for Programming Language 1, which is a high-level programming language designed for applications requiring the preparation of computer programs for most business and scientific functions. PL/1 is intended to be a language which takes advantage of the strong and weak points of FORTRAN and the business-oriented languages, and bridges the gap between them.

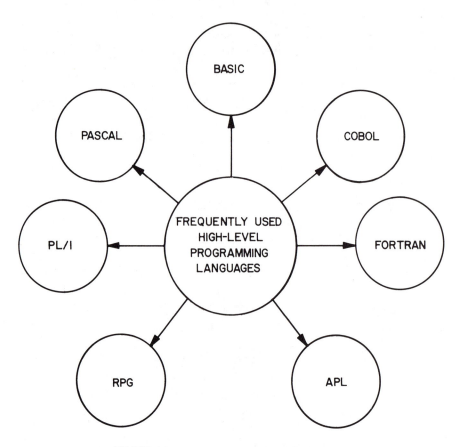

FIGURE 1-11 High-level programming languages.

PASCAL is a high-level programming language named after the famous French philosopher and mathematician Blaise Pascal. One of the most common applications of PASCAL is in the teaching of computer programming. It is a relatively simple language and easy to learn. Consequently, it is used as a tool for teaching people how to write programs.

FIRMWARE OR MICROCODE

You know that computer hardware consists of the computer itself and the various peripheral devices used for interacting with the computer in inputting and outputting data. You also know that software consists principally of the computer programs used to direct the computer in performing the various tasks required of it. There is also a third concept which falls between hardware and software.

This third component is called "firmware" or microcode. Microcode is a sequence of commands built into the computer and executed automatically by the hardware. Microcode is stored in a special read-only memory section so that it cannot be altered by a user. This is what read-only memory, or ROM, means. Instructions can be read from it but not written to it. Microcode gives manufacturers of computers the ability to tailor computers to specific applications without making major changes to the hardware.

COMPUTER OPERATION

There are three stages of operation in a digital computer: input, processing, and output. Input is any technique used to communicate with the processor in entering data into a computer. Output is any technique used for communicating with the processor to get data out of the computer in the desired form. Processing involves reading, manipulating, or performing any other type of operation on data in accordance with the directions contained in a computer program.

In order for there to be input to and output from the processor, there must be a means of control and a means of data flow. Control is accomplished by an input/output control system, or IOCS. An IOCS is a standard set of routines designed to initiate and control input and output operations in a computer system. The actual pathway for data flow is called an input/output channel.

An input/output channel is a hardware pathway that allows independent communication between a storage unit and any number of input or output devices. In a typical computer application setting, most data will be entered (or input) using a keyboard, output in soft form on a terminal display, and output in hard form on a printer.

SUMMARY

A computer is an electronic machine that has storage, logic, and mathematical capabilities and is able to perform certain tasks at extremely fast speeds. Humans and computers have compensating strengths and weaknesses which make the human/computer team a particularly effective one. The two main classifications of computers are digital and analog. Digital computers are the type used in CAD/CAM applications.

Computers that fit the definition presented above had their beginnings in the 1940s. The first computer that began to fit this definition was the Mark I computer. The various generations of computers since the Mark I have been characterized by some specific technological development in the field of electronics. The first generation of computers was characterized by the vacuum tube; the second generation was characterized by transistors; and the third generation was characterized by the integrated circuit. We are currently in the third generation of computers.

The machines and equipment in a computer system are called hardware. Hardware can be divided into four broad categories: the processing unit, secondary storage devices, input devices, and output devices.

The main memory in a computer is classified as an 8-, 16-, 32-, 48-, or 64-bit memory. These classifications refer to the size of the computer word the memory can handle. A computer word is the amount of data a computer can retrieve from memory per instruction. A bit is a binary digit represented in the computer by electrical impulses that are either on or off. The memory size in a computer is usually stated in kilobytes, megabytes, or some higher prefix of bytes. One kilobyte means 1,024 bytes of memory. A byte is generally considered to be 8 bits. Other important features of computer memory are memory mapping, parity memory, cache memory, and virtual memory.

The main memory of a computer is actually directly connected to the central processing unit and is usually housed in the same console as the processing unit. Main memory can be divided into three categories: main data storage, control storage, and local storage. Secondary storage is any storage medium outside of the main memory, such as magnetic tape and floppy disks. Secondary, or auxiliary, storage can be divided into two broad categories: sequential access auxiliary storage and direct access auxiliary storage.

There are a number of different ways of representing computer data. The most common method is the binary numbering system. In the binary numbering system, electrical impulses which represent data in the computer are represented themselves by ones or zeros. Any number can be represented in the binary numbering system by a series of ones and zeros. Through the use of special binary coding schemes—such as the ASCII coding scheme, the EBCDIC coding scheme, and the BCD coding scheme—the binary system can be used for representing letters and special characters in addition to numbers.

Computer software refers to computer programs, documentation of programs, and the various types of technical and reference manuals that go with computer systems. However, when most people use the term "software," they are talking strictly about computer programs. A computer program is a specially coded set of instructions that directs the computer in performing all operations. Some frequently used computer programming languages are machine language, assembly languages, BASIC, COBOL, FORTRAN, APL, RPG, PL/1, and PASCAL.

Firmware is another name for built-in microcode in a computer. Microcode is a sequence of commands built into the computer and executed automatically by the hardware. It is stored in a special read-only memory (ROM) section so that it cannot be altered by the user.

There are three phases of operation with any computer: input, processing, and output.

Chapter One REVIEW

1. Define the term "computer."
2. List four characteristics which distinguish computers from other machines.
3. Give three examples of offsetting strengths and weaknesses which computers and humans have.
4. Differentiate between digital and analog computers.
5. Give the technological developments which have characterized each of the three generations of computers to date.
6. What was the first computer to be commercially used, and who developed it?
7. Explain the computer memory related to terms "bit," "byte," and "word."
8. Explain the following terms:
   RAM
   Memory Mapping
   Parity Memory
   Cache Memory
   Virtual Memory
9. What are the three categories of main memory?
10. Explain the two different categories of secondary storage.
11. What is a "computer program"?
12. What are the three phases of operation of a digital computer?

In Chapter 1, you learned all of the generic information about computers that manufacturing students need in order to understand CIM. This chapter takes the next step in helping you learn how computers are used in a variety of manufacturing settings. With this, you will have developed the solid foundation of computer knowledge that all students of CIM need.

Major Topics Covered

- Microcomputers
- Minicomputers
- Mainframe Computers
- Supercomputers
- Computers in Design, Drafting, and Manufacturing

Chapter Two

Computers in Design, Drafting, and Manufacturing

case study

Increase in complexity and density of electronic packaging combined with markets segmented by demand and competition demand that designers and engineers have the necessary tools to bring products profitably to market in record time. Development cycles of one or two years may have been tolerable as little as five years ago; but today, with such factors as rapid changes in technology, imitators, and multiple technological solutions, a one- or two-year cycle cuts deeply into the potential profitability of any product. *In a study by McKinsey & Company, it was shown that shipping a product six months late has a much greater impact on its total profits than overrunning its development cost by 50% or pricing it 9% too high.* In meeting both their immediate needs and in preparation for perceived future development requirements, larger companies have invested heavily over the years in large turnkey solutions. Spurred by the success of CAD/CAM systems individually constructed by the aerospace manufacturers in the 1960s, large corporations spent big sums to bring CAD into their companies. These high-speed and performance systems were umbilically linked to mainframes in the beginning, and later to minicomputers. From this eventually evolved the 16- and 32-bit workstations. Although these offered practical solutions for their time and purpose, the very nature of these "tools" did much to carry the power and performance down to the individual level. Smaller companies with fleetness of foot and the ability both to identify new markets and to penetrate the bigger, more established ones were able to thrive without the need for these expensive and cumbersome systems.

Enter the PC in the 1980s and everything changed. More and more, computing power, performance, main memory size, disk memory size, and software closed the gap between PCs and mainframes. This allowed also for this valuable resource to be passed down to the individual engineer, whereas the mainframe was a shared resource. Notwithstanding the computer itself—whether a mainframe, a workstation, or a PC—the one major distinction in program offerings that will have the most singular impact on the ability to bring products to market faster is 3-D. To me, a CAD system without 3-dimensional capability is like an automobile

Courtesy of Micro-Control Systems, "An Applications Story" by Harold L. Bowers, 1986.

A Discussion of the Relative Merits
of a 3-Dimensional, PC-Based CAD System
vs. Other Systems

with no reverse gear. You know where you want to go, but there are a lot of dead-end streets. The big players—ComputerVision, IBM, Autotrol, and McDonnell Douglas—have excellent 3-D capability, but they also have a high pricetag. Today, while a PC CAD system can cost under $10,000, a state-of-the-art 32-bit system can easily run $100,000 to $200,000. The big question here is: *Given you can find a PC system with "comparable" abilities, is the extra capability worth the extra cost?* This, of course, depends on the application. Where sizable computing ability is required, the big money machines will, for the time being, be economically feasible. But the large share of duties that takes up the engineer's and designer's time and effort could be satisfied with a PC-based system. This did indeed provide for a ready and willing market. Even though early PC CAD/CAM programs and workstations had only a 2-D capability, the potential for productivity improvements was significant and sizable. But one should not overlook the fact that the ability to build an effective 3-D database at the conceptual stage of product development carries with it all the elements of potential market success, such as decreased number of prototype cycles; fast, interactive design changes; producibility analysis; computerized testing; cogenerated technical documentation; computer linked to automated manufacturing machines; and all the other de facto and well-established attributes that are associated with the use of 3-D databases and sorely missed by even the most elaborate 2- or 2½-D system. In effect, 3-D capability means getting the product to market within schedule, under cost, profitably priced, and with sufficient quality.

One of my first experiences with CAD was on a ComputerVision 4X system where I was responsible for the mechanical engineering design and development for a series of advanced, handheld, 2-way radios. Starting with the industrial design and ending with the accessories, this system was utilized in all its capacity to model, test, fit, analyze, and prototype an entire series of new products. Owing to the very nature of these products, it was necessary to design them using all the 3-D capabilities the system could offer. Extremely dense packaging required such features as surface mount components, ergometric design (must fit com-

(continued)

fortably in the customer's hand and be easily operated by either hand), portability, and high reliability—smaller and more features than had ever been offered by this manufacturer in such a product line. Without this CAD/CAM system, it would have been extremely difficult and time-consuming (if not impossible) to integrate all these product specifications into such a design in any reasonable development time cycle. Within six months, models were being made from computerized tapes to prototype milling machines for presentation to the top company officials. Needless to say, this would have consumed many more man months of effort had we either done it manually or used only the 2-D (e.g., drafting) capability of the ComputerVision system.

When I later joined a small company involved in the design of electronic communications equipment, I immediately recognized the need for this same type of ability. However, this time I was without the large company budget or skill resources. This, obviously, ruled out the purchase of a large mainframe-based system. Also, severely constricted delivery schedules for equipment meeting military specifications demanded not only improvements to drafting productivity but acceleration of the actual creation of the products. As the IBM PC AT had just been introduced, CAD drafting programs were coming rapidly on the scene and were readily available. While this would help alleviate the obvious bottleneck of drawing production, the design stage and all the other ancillary functions (such as technical documentation, redesign, modeling, tolerance analysis, and producibility) would have to be performed in a somewhat serial fashion, with the data manually "reentered" for each function. Realizing that these 2- and 2½-D systems would only serve as a stopgap, I began to hunt in earnest for a system that could meet the needs of this growing company, but on a shoestring budget. My search ended when I happened to find a small software company in Connecticut that had a 3-D system that looked on paper to be somewhat promising. As I was to be in the area on other business the next week, I called and asked if I could come in for a demonstration. They said they would be very happy to show me the capabilities of their software, as they had it set up on a number of different PC systems at their facility and I could readily see how it operated in a "real" PC. At that demonstration I found not only a 3-D CAD/CAD system but the rudiments of a much larger and more expensive turnkey system. With this software, the engineer/designer can design a part of three axes while rotating it to see it from every view, model it as a solid or wireframe, and "explode" it to create exact replicas suitable for technical documentation or "critical" review. As it had to appeal to PC users, it was surprisingly user friendly and offered an approach that was obviously designed around the very meaning of 3-D and not simply a 2- or 2½-D system modified to give 3-D characteristics. Well integrated into this software was a very effective and useful drafting utility that could be dimensioned to ANSI and ISO drafting standards that allowed quick and effective downloading of designs to useful drawings. These could then immediately be put on the street for bids or production. The CADKEY developers had apparently devised a way in which both designers and draftsmen could get maximum utility out of a very complex, combined 2-D and 3-D system well suited to product design and development while staying well within the PC price range. Makers of 2- and 2½-D PC-based CAD/CAM software, while making productivity improvements on the order of ten to fifteen times, were, in my opinion, downplaying the real gains to be made in this area. These gains were to be achieved by the development of highly flexible and interactive 3-D databases tied to an explosion of labor and capital savings devices such as numerical controlled machines, plastic injection molders, database management systems, technical documentation development and printing, and computerized testing. In the area of mechanical/electronic packaging, a system such as a PC-AT, or one compatible with CAD software (e.g., CADKEY), properly constructed, properly managed, and with the proper skill levels "creating" products, could be a very effective and dynamic tool in shortening the time to market. Ultimately, it could provide the basis for ensuring such elementary market forces as giving the customer a well-built, quality product for a profitable price.

Computers come in four sizes. From smallest to largest, they are: the microcomputer, the minicomputer, the mainframe computer, and the supercomputer (Figure 2-1). Although most computers used in design, drafting, and manufacturing applications are minicomputers and microcomputers, there are mainframe-based CAD/CAM systems. A supercomputer would not normally be used in a design, drafting, or manufacturing setting. However, students of CAD/CAM should be familiar with the supercomputer and some of the applications in which it is used.

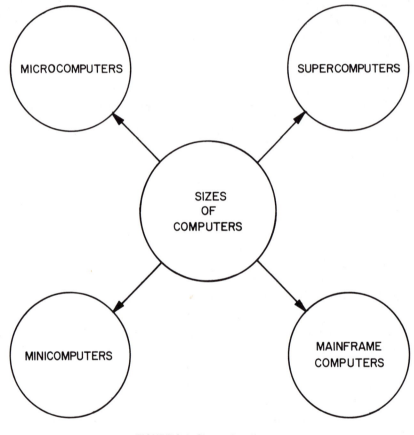

FIGURE 2-1 Sizes of computers.

MICROCOMPUTERS

A microcomputer is the smallest type of computer. It is a complete computer on a single printed circuit board. Other names for the microcomputer are desktop computer, home computer, personal computer, portable computer, and small business computer. Figure 2-2 is an example of a popular microcomputer.

FIGURE 2-2 AT&T PC 6300 Plus microcomputer. Courtesy of AT&T.

Microcomputers are used extensively in CAD/CAM settings now, but this was not always the case. In the late 1970s, microcomputer technology had not advanced to the state where microcomputers had the memory capabilities, processing speed, and graphic capabilities necessary to function in a CAD/CAM setting. Early microcomputers were used principally for such applications as pocket computers, home computers, computers for game playing, and word processing.

Microcomputer technology really came of age in the early 1980s, when IBM decided to enter the market with its IBM PC, AT, and XT models. Since that time, numerous other vendors have rushed to enter the market with their own products. However, IBM's microcomputers have become the yardstick against which other microcomputers are measured. Many of the microcomputers which have followed are marketed as IBM PC compatible or IBM PC "Work-a-likes." It is estimated that IBM personal computers are used in over 40% of the CAD/CAM applications where microcomputers are used. The ten leading vendors of microcomputers used in CAD/CAM applications are:

IBM
Apple
DEC
Texas Instruments
Hewlett-Packard
Compaq
AT&T
NEC
Victor
Columbia

Figure 2-3 is an example of a CAD system based on the IBM PC AT microcomputer.

MINICOMPUTERS

There really is no clear-cut definition for the minicomputer. The only way that a minicomputer can really be defined is by looking at it in relative terms. A minicomputer is a scaled-down version of a full-scale computer. A minicomputer would generally have less memory and, in some instances, a smaller word size than a full-size computer, but these criteria do not always apply. This is because a modern minicomputer may have more memory and a larger word size than an older full-size computer.

Another way that minicomputers have been defined is as any computer, other than a microcomputer, costing less than some set amount.

FIGURE 2-3 IBM-PC/XT based CAD system. Courtesy of Mentor Graphics Corporation.

That amount has been set at $50,000, $60,000, and even $75,000 by some vendors. Figure 2-4 is an example of a minicomputer—the Burroughs XE-550.

On the exterior, a minicomputer is as likely to resemble a microcomputer as a mainframe. However, the inside architecture of a minicomputer is the same as that of a mainframe computer. Minicomputers fill a void between what microcomputers are able to do and what mainframe computers are able to do. There are users whose needs do not require

FIGURE 2-4 Burroughs XE-550 superminicomputer can be used as a mainframe computer.
Courtesy of Burroughs Corporation.

the capabilities of a mainframe, yet exceed the capabilities of a micro-computer. This is what led to the development of the minicomputer.

There are a number of vendors involved in the manufacture and marketing of minicomputers. Some leading minicomputer vendors are:

- Digital Equipment Corporation (DEC)
- Data General

- Hewlett-Packard
- IBM
- Prime
- Honeywell Information Systems
- Wang Laboratories

MAINFRAME COMPUTERS

Full-scale computers have come to be known as "mainframe" computers. Again, there is no set definition for the term mainframe computer. Mainframe computers are not the largest computers available, but they are the largest commonly used computers. Supercomputers are, of course, larger than mainframes. However, supercomputers are not widely or commonly used.

If one disregards technological developments which blur the difference between minicomputers and mainframe computers, especially when the minicomputer is a new model and is compared against an older mainframe, there are some general rules which apply. Mainframe computers are generally more expensive, have more memory, have a faster processing speed, and have a word size of 16 bits or higher. A typical mainframe computer has a 32-bit word size. Figure 2-5 is the Burroughs V380 Series mainframe computer.

There are a number of different vendors who manufacture and market mainframe computers. Some of these, as you will see, also manufacture and market minicomputers. The leading vendors of mainframe computers are:

- Burroughs
- Sperry
- Hewlett-Packard
- Digital Equipment Corporation (DEC)
- Honeywell
- IBM

SUPERCOMPUTERS

There are some computer applications which need more memory, a faster processing speed, and a larger word size than even large full-scale mainframe computers can provide. Such applications include large research labs, the National Weather Bureau, the Internal Revenue Service, large airlines, large oil companies, and large utility companies.

FIGURE 2-5 Burroughs V380 series mainframe computer. Courtesy of Burroughs Corporation.

In the mid-1970s, in response to the need for computers exceeding the capabilities of large mainframe computers, supercomputer technology came into being. Typically, a supercomputer would have such characteristics as an input/output rate that exceeds 3 billion bits per second, as many as 4 million 64-bit words of primary storage, and the ability to process over 50 million instructions per second. The leading vendors of supercomputers are Cray Research and Control Data Corporation.

Microcomputers, minicomputers, and mainframe computers are manufactured on a normal production or assembly-line basis. In other words, they are mass-produced and distributed to the market. This is not true of supercomputers. Supercomputers are normally built one at a time, for a specific buyer, in a specific application. It typically takes from six months to a year to build one supercomputer.

COMPUTERS IN DESIGN, DRAFTING, AND MANUFACTURING

Computers are used in a variety of ways in design, drafting, and manufacturing. Some of the most common applications of computers in CAD/CAM are:

1. as the nucleus of computer-aided design and drafting, or CADD systems (Figure 2-6);
2. for programming and controlling numerical control manufacturing machines;
3. for programming and controlling industrial robots (Figure 2-7); and
4. as the functional nucleus of computer-integrated manufacturing, or CIM systems, and flexible manufacturing systems, or FMS.

Another computer-related device frequently used in CAD/CAM settings is the programmable controller. Students of CAD/CAM should be familiar with the programmable controller, how it is used in manufacturing settings, and the advantages it offers over traditional control devices.

The Programmable Controller

Programmable controllers came about in the late 1960s as a technological replacement for the traditional relay logic systems used in the control of production equipment in manfacturing. A programmable controller is an electronic device which generates output signals according to the logic operations it performs on input signals. The input, logic operations, and resulting output of a programmable controller are the result of instructions contained in the program.

Programmable controllers serve the same purpose that wired relay panels have always served, and they accomplish this purpose in a similar manner. However, programmable controllers offer a number of advantages over the traditional relay systems. Some of the more important of these are:

1. As the name implies, a programmable controller can be programmed. This is an important advantage, because it makes

FIGURE 2-6 ICON CADD system. Courtesy of Summa Technologies.

changing inputs, the logic operations that are performed, and the resulting outputs much easier. Changing a program is a much faster and easier process than rewiring a relay panel.

2. Because programmable controllers rely on miniaturized integrated circuits, just as computers do, they tend to be smaller than the traditional wired control panels. Consequently, they require less space.

3. Interfacing a programmable controller with computer systems available in CAD/CAM applications is much easier than trying to interface the traditional wired relay panels.

Programmable controllers are not computers in the strictest sense of the word. However, as the technology associated with computers

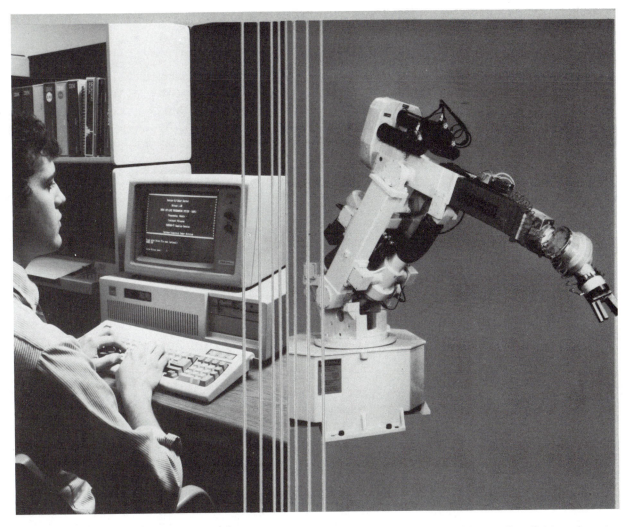

FIGURE 2-7 Using a microcomputer to program an industrial robot off-line. Courtesy of Cincinnati Milacron. (Note: Safety equipment may have been removed or opened to clearly illustrate products and must be in place prior to operation.)

and programmable controllers has advanced over the years, the two devices have become more closely related.

Programmable Controller Components

There are several different makes and models of programmable controllers produced by a variety of vendors. Regardless of the make, model, or vendor, however, all programmable controllers have five basic components.

1. Processor
2. Memory

3. Programming device
4. Input/Output interface mechanisms
5. A power supply

Figure 2-8 is a schematic representation of the components of a programmable controller. The memory and central processing unit of a programmable controller serve the same purposes as the memory and central processing unit of a computer. Programming devices for writing the programs which direct the operations of the programmable controller are of two types.

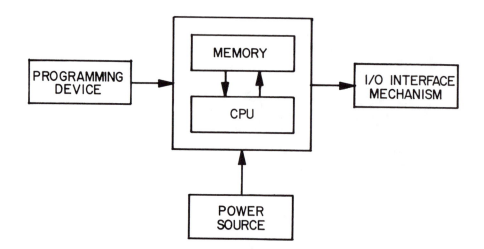

FIGURE 2-8 Components of a programmable controller.

The first type is the CRT terminal with a keyboard. This is the most convenient and most popular type of programming device. The second type of programming device is a small handheld keybox or keyboard device. The input/output interface mechanisms allow the central processing unit to be connected to input/output devices. The power source takes in electricity in the form of alternating current (AC) and converts it to the necessary direct current (DC).

One of the factors which has made programmable controllers so popular is that they are almost immediately accepted by the production people who must use them. This is because experienced manufacturing personnel who are already familiar with how to use such things as relay ladder diagrams in setting up wired relay panels do not have to learn anything new. The knowledge they already have is all that is necessary in order to be able to write programs for a programmable controller.

There are a number of different programming formats which can be used in writing instructions for programmable controllers. Some vendors have their own programming formats for their equipment. However, if a manufacturing person is familiar with relay ladder diagrams, such as were used in setting up wired relay panels, he or she should have no problem making the transition to programmable controllers.

In spite of their similarities, there are several important differences between programmable controllers and computers. CAD/CAM students should be familiar with these differences. The most important differences between the two devices are not technological. Rather, they are to be found in how they are used. Programmable controllers are designed specifically for use in controlling the operations of production equipment. Because of this, they are physically structured in slightly different ways than computers.

One such way is in the interfacing mechanisms built in so that programmable controllers can be tied directly to production equipment. No such interfaces exist in the typical minicomputer or microcomputer. Computers designed to serve a variety of purposes in a variety of settings. Consequently, special provisions must be made in order for them to be interfaced with production equipment.

A second difference is in how the two devices are programmed and who is able to program them. The languages used in programming a programmable controller are vastly different from those used in programming a computer. Figures 2-9 and 2-10 are examples of commonly used computer languages. Figure 2-9 is a sample COBOL program. Figure 2-10 is a sample Fortran program. Figure 2-11 is an example of a program for a programmable controller and a close-up of a keyboard showing what keys would be used in keying this particular program. From Figures 2-9, 2-10, and 2-11, one can see that computer programs and those for programmable controllers are radically different. In addition, programmable controllers are programmed by the manufacturing personnel who use them. This is not typically true of computers. Very few computer users actually program computers.

A final difference is in how computers and programmable controllers are packaged. Computers are packaged in such a way as to be physically user friendly and aesthetically pleasing in an office environment. Programmable controllers, on the other hand, are packaged in such a way as to withstand the rigors of the production environment. Consequently, they look much different.

```
001900 REMARKS.  WR 3985 INITIALIZES S-COUNSELOR TO SPACE.
002000 ENVIRONMENT DIVISION.
002100 CONFIGURATION SECTION.
002200 SOURCE-COMPUTER.   IBM-370.
002300 OBJECT-COMPUTER.   IBM-370-138.
002400 SPECIAL NAMES.
002500     C01 IS TOPLINE.
002600 INPUT-OUTPUT SECTION.
002700 FILE-CONTROL.
002800     SELECT  STUBIO-MASTER COPY ASTUBD.
002900     SELECT    LISTING   COPY   ASPRNT.
003000 DATA DIVISION.
003100 FILE SECTION.
003200 FD  STUBIO-MASTER COPY STUBIO.
003300 FD  LISTING
003400     LABEL RECORDS ARE OMITTED.
003500 01 LIST-RECORD         PIC X(132).
003600 WORKING-STORAGE     SECTION.
003700 77 W-S-START     COPY   WSSTRT.
003800 77 RECORD-CCUNT        PIC 9(5) VALUE O.
003900 77 WRITE-COUNT         PIC 9(5) VALUE O.
004000 01 SPEC-CARD.
004100     05 SPEC-CONTROL     PIC X.
004200     05 SPEC-COUNT       PIC (9)6.
004300 01 FILE-STATUS     COPY   FLSTAT.
004400 PROCEDURE DIVISION.
004500     ACCEPT SPEC-CARD.
004600     IF  SPEC-CONTROL = 'U'
004700           OPEN I-O   STUBIO-MASTER
004800     ELSE    OPEN   INPUT   STUBIO-MASTER.
004900     OPEN OUTPUT LISTING.
005000 READ-STUBIO.
005100     READ STUBIO-MASTER NEXT AT END
005200           GO TO END-OF-JOB.
005300     ADD 1 TO RECORD-CCUNT.
005400     IF RECORD-COUNT > SPEC-COUNT GO TO END OF JOB.
005450     IF S-COUNSELOR = SPACE GO TO READ-STUBIO.
005500     MOVE ' ' TO S-COUNSELOR.
005600     IF SPEC-CONTROL = 'U'
005700        PERFORM REWRITE-STUBIO THRU REWRITE-EXIT.
005800     GO TO READ-STUBIO.
005900 REWRITE -STUBIO.
006000     ADD 1 TO WRITE-COUNT.
006100     REWRITE  BIO-RECORD
006200        INVALID KEY  DISPLAY ' WRITE ERROR ON STUBIO'
006300           STOP 05   STOP RUN.
006400 REWRITE-EXIT.   EXIT.
006500 END-OF-JOB.
006600     CLOSE  STUBIO-MASTER.
006700     IF SPEC-CONTROL = 'U'
006800        DISPLAY 'RECORDS SELECTED =>>>   ' RECORD-COUNT
006900        DISPLAY 'RECORDS WRITTEN =>>>   ' WRITE-COUNT
007000        DISPLAY ' STUBIO-MASTER UPDATED
007100     IF SPEC-CONTROL = 'E'
007200        DISPLAY 'RECORDS SELECTED =>>>   ' RECORD-COUNT
```

FIGURE 2-9 Portion of a sample COBOL program.

```
    IMPLICIT   INTEGER*2(I-N).REAL*4(A-H,O-$)
    COMMON    Z(4).A(4.8),V(4.8),S(4.8),H(4.8),W(4),P(4.8),Y(4.8),
  I           C(4.8),Q(4.8),E(4.8),F(4.8),R(4.8),D(4.8),X(4.8).
  I           J1(4.8),I1(4.8),N(4.8),O(4.8),G(4.8),U(4.8),K(4.8)
    REAL F1,PO,A9
    REAL MO,N1.J1,I1,K3,N2,N3
    REAL HO,BO,A0,00,01,02,Q5
    REAL N
    INTEGER T,Z,W
    DATA  HO,BO,MO,A0,00,01/18.E3,.30,L.45,2.2,4.0,1.5/
    DATA 02,Q5,K1,K2,K3 /.012,1.1,.15,5.E3,.3/
  1 FORMAT ('1' //30X,'TIME = ',I2)
 16 FORMAT(//30X, 'DECISION IS',14.5X, 8HFOR FIRM.12)
 18 FORMAT( 30X,4F10.2)
856 FORMAT(// 30X, 'INPUT DATA' /)
851 FORMAT(//30X ,30HPROFIT AND LOSS STATEMENT,FIRM, 12)
852 FORMAT(// 30X,'NUMBERS IN THOUSANDS OF DOLLARS')
855 FORMAT(/// 30X,'SALES',F29.1)
866 FORMAT(//30X, 'VARIABLE COSTS',F21.0)
871 FORMAT(// 30X, 'FIXED COSTS',F21.0)
879 FORMAT(// 30X,'VARIABLE COSTS,TESTMKT',F13.1)
883 FORMAT(// 30X,'FIXED COSTS,TESTMKT',F15.1)
885 FORMAT(// 30X,'ADVERTISING',F21.0)
886 FORMAT(// 30X,'CHANNEL PROMOTION',F15.0)
888 FORMAT(1X, 4F10.2)
890 FORMAT(// 30X,'PROFITS',F24.1)
990 FORMAT(// 30X,'MKT RESEARCH FINDINGS,FIRM' ,12,2X,6HPERIOD,12)
991 FORMAT(// 30X,'AVAILABILITY', F23.2)
992 FORMAT(// 30X,'INITIAL PURCHASE', F19.2)
993 FORMAT(// 30X,'RATE OF REPURCHASE' ,F17.2)
994 FORMAT(// 30X,'CONSUMPTION RATE',F29.2)
1011 FORMAT( 30X,4F10.2)
    DO 11 I=1,4
    DO 11 J=1,8
    A(I,J) = 0.0
    R(I,J) = 0.0
    S(I,J) = 0.0
    U(I,J) = 0.0
    G(I,J) = 0.0
    J1(I,J) = 0.0
 11 CONTINUE
    READ (1,15)  T
    WRITE (3,1) T
 15 FORMAT(12)
854 FORMAT (1H1)
    DO 20 I=1,4
    READ (1,15) Z(I)
    WRITE (3,16) Z(I),I
 20 CONTINUE
    WRITE (3,856)
    IF(T .EQ. 1) GO TO 115
5700 FORMAT(/ 30X, 'HISTORY DATA')
    WRITE(3,5700)
    IF(T-8)5000,5000,5500
5500 WRITE(3,5600)T
```

FIGURE 2-10 Portion of a sample FORTRAN program.

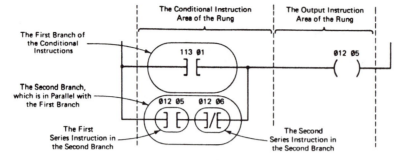

The Conditional Instruction Area of the Rung

The Output Instruction Area of the Rung

The First Branch of the Conditional Instructions

113 01

012 05

The Second Branch, which is in Parallel with the First Branch

012 05   012 06

The First Series Instruction in the Second Branch

The Second Series Instruction in the Second Branch

(a)

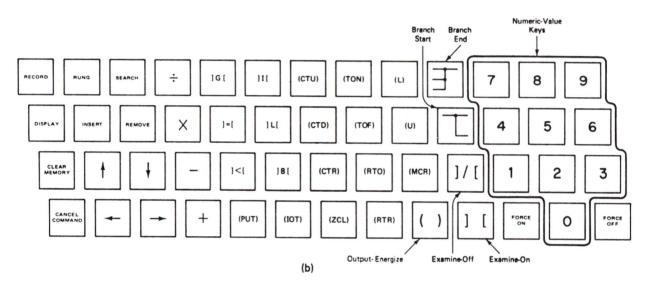

(b)

FIGURE 2-11 Sample program (instruction rung) for a programmable controller. (From INDUSTRIAL SOLID-STATE ELECTRONICS: DEVICES AND SYSTEMS, 2/E, © 1986, p. 100. Reprinted by permission of Prentice-Hall, Inc., Englewood Cliffs, New Jersey.

SUMMARY

Computers come in four sizes. From smallest to largest they are microcomputers, minicomputers, mainframe computers, and supercomputers. A microcomputer is a complete computer on a single printed circuit board. There is no clear-cut distinction between the minicomputer and mainframe computer. However, in general terms, a mainframe will usually have more memory, a larger word size, and a faster processing speed than a minicomputer. Supercomputers are especially designed for those few applications which require more memory, a larger word size, and faster processing speed than even a mainframe computer can offer.

Computers are used in CAD/CAM settings for programming and controlled numerical control machines, programming and controlling industrial robots, and as the functional nucleus of CADD, CIM, and FMS systems.

Programmable controllers are special devices which were developed in the 1960s to replace the traditional relay logic systems used in controlling production equipment. Programmable controllers are not actually computers in the strictest sense of the word. However, the two devices become more closely related with every technological advance that occurs in each area.

A programmable controller, regardless of make, model, or vendor, has the following components: processor, memory, programming device, input/output interface mechanism, and a power supply.

Chapter Two REVIEW

1. Define the term "microcomputer."
2. What are several other names for the microcomputer?
3. What was the most important event in the coming of age of the microcomputer in CAD/CAM applications?
4. Name five leading vendors of microcomputers used in CAD/CAM applications.
5. If forced to do so, how would you distinguish between a minicomputer and a mainframe computer?
6. Name three leading vendors of minicomputers.
7. Name three leading vendors of mainframe computers.
8. What is a supercomputer?
9. List four applications of computers in CAD/CAM.
10. What is a programmable controller?
11. What are the components of a programmable controller?

Automation of the design and drafting processes began to see wide-scale implementation in 1979 in the form of computer-aided design and drafting (CADD). When full integration of manufacturing has been accomplished, CADD will be one component in an overall CIM system.

Major Topics Covered

- What Is Design?
- What Is Drafting?
- What Is CADD?
- Advantages of CADD
- Complaints About CADD
- Applications of Computers and Software in Design and Drafting
- CADD Hardware
- CADD System Operation
- CAD/CAM Integration and Networking
- The CADD Market
- CADD-Related Math

Chapter Three

Computer-Aided Design and Drafting

Industrial Acoustics Co., Inc., a well-known Bronx, New York, manufacturer of acoustical enclosures, silencers for the aircraft and air-conditioning industries, and other noise-control products, recently began a pilot installation of a new CAD/CAM technology. Involving CADKEY (a 3-D PC-based CADD system) and CAD Punch™ (a sheet-metal fold-out system), Industrial Acoustics is the first company to use this new fabrication system.

Formed sheet metal is a major component of most of the products sold by Industrial Acoustics. The formed part shown here is a door for a soundproof room. The door is defined and designed with CADKEY, the front end system for CADPUNCH. As x, y, and z data is transferred from CADKEY to CADPUNCH, the program "unbends" the design, adjusts for bend factors, material adjustments, and so on (without user intervention), and produces a flat blank of the design. After adding the sheet size and clamp locations visually in CADKEY, CADPUNCH uses the tool load present in Industrial Acoustics Trumpf punch press to *automatically* produce an N/C tape file and a plot of all punch fits directly from the sheet drawing for verification.

Previous to this system, the company employed 2-D manual techniques in constructing its designs, then would have to calculate out the bend factors, then manually draw out the flat blank. The dimensions would have to be done again, for the blank typically would not have the same dimensions as the 3-D design because of the bend allowances. The user would then take the flat blank and input the data all over again into the computer-aided programming system for the trumpf press.

This new process, going from 3-D design to flat sheet layout to a punch press, greatly facilitates the design process and alleviates countless manual calculations and chances of error. The CADKEY program is accurate to 14 digits of accuracy, so the user need not worry about the initial information being input into the CAD Punch system. Better communication between the design/drafting and manufacturing departments is a necessary and beneficial by-product.

Courtesy of Micro-Control Systems, Inc., "3-D CADD-CAM."

The term "CAD" is alternately used to mean Computer-Aided Design or Computer-Aided Drafting. To clear up the confusion this can cause, the term "CADD" or Computer-Aided Design and Drafting can be used. This term properly depicts the relationship of the two separate but intertwined concepts.

WHAT IS DESIGN?

Design involves applying scientific principles to the solution of everyday problems in a step-by-step systematic manner. In general terms, the design process is five steps through which scientific principles are applied to the solution of everyday problems (Figure 3-1).

1. Define the problem.
2. Conceptualize and analyze the design.
3. Test and optimize the design.
4. Document the design.
5. Move the design to production.

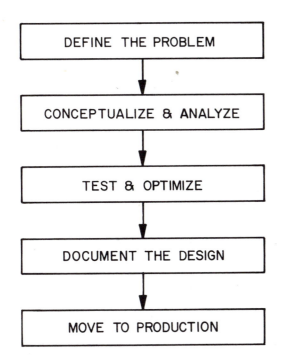

FIGURE 3-1 The design process.

The design process is undertaken after a specific problem or need has been identified. The identification of the problem occurs in the research and development phase of manufacturing. Once scientists and engineers working in research and development have identified the need for a product or discovered a problem, the design process is used to make the new product a reality or to solve the problem.

Define the Problem

In this phase, the problem is examined from every possible perspective to ensure that it has been clearly defined before further work takes place. Clear and definitive specifications are written for the design. These specifications would include all physical, operational, functional, and cost parameters for the design.

Clearly defining the problem before further work on a design proceeds is critical. Few designs are accomplished by a single engineer. Most designs require the cooperative work of a team of designers and engineers. Consequently, it is important for all parties involved to understand exactly what the problem is before they begin to work on solutions. Much time, money, and energy can be wasted if all members of the team are not working toward the same end.

Conceptualize and Analyze

In this phase, designers and engineers develop a concept of the overall design and each subunit of it. The design as a whole and each functional part of it are carefully analyzed. Problems discovered during analysis are corrected during redesign, or are "designed out." During this phase, the design is documented through preliminary drawings, sketches, calculations, written reports, preliminary parts lists, and preliminary bills of material. Once the conceptualization of the design and analysis of it are complete, the design is ready for testing and optimization.

Test and Optimize the Design

A fully conceptualized and analyzed design represents the best effort of the designer and engineer on paper. In this step, the on-paper design must be tested in a more realistic mode. This might involve the building and testing of prototypes, live scale models, computer models, or a combination of all of these. A prototype is a life-size nonproduction version of the actual design. Prototypes are tested in a variety of conditions and circumstances to ensure that the design will actually perform as required. Scale models are tested in manners similar to prototypes. Computer models are tested through computer controlled simulation.

Regardless of how a design concept is tested, the testing step gives designers the opportunity to optimize the design by further refining it and correcting "bugs" found in it. Often, a test on a live prototype or model will reveal problems not readily apparent on paper. Once a design has been fully tested and optimized, it must be fully documented so that it can be produced in the desired quantities.

Document the Design

In this phase, all preliminary drawings, sketches, bills of material, parts lists, and other forms of documentation are transformed into a finished documentation package. Such a package will include complete working drawings, parts lists, bills of materials, written specifications, and all calculations necessary to verify all aspects of the design.

In the past, designs have been documented manually, using such tools as drafting machines, parallel bars, triangles, mechanical pens and pencils, scales, and so forth. In a modern design setting, documentation is accomplished on a CADD system (Figure 3-2).

Once the design has been completely and accurately documented, the documentation package is forwarded to production so the design can be manufactured.

Move to Production

In the past, the documentation package was forwarded to production personnel in hard-copy form. This means that they received blueprints of the drawings, and copies of specifications, bills of material, and parts lists. In a modern CAD/CAM setting, the documentation package can be forwarded electronically through a Local Area Network (LAN), as you will see later in this chapter and in other chapters which follow.

WHAT IS DRAFTING?

Drafting is a very important part of the design process. The simplest definition of drafting is that it is documenting the design process. Such documentation is critical. Designers and engineers conceptualize a design in their minds. Through drafting, these concepts and ideas are

FIGURE 3-2 A modern CADD system. Courtesy of CalComp, a Sanders Company.

converted into graphic form so that all parties involved are able to actually see what designers and engineers have in mind.

Any form of graphic communication used to document the design process falls under the heading of drafting. Sketches, preliminary drawings, complete working drawings, parts lists, and bills of materials are all products of drafting because they all document, in some way, the design process. Figure 3-3 is an example of an item of design documentation produced through drafting.

WHAT IS COMPUTER-AIDED DESIGN AND DRAFTING, OR CADD?

You know that design involves applying scientific principles to the solution of everyday problems. You also know that drafting involves

Mechanical Drawing

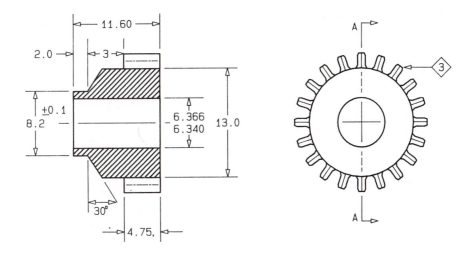

Section View   Top View

Plotted on the Hewlett-Packard 7550 Graphics Plotter using the IBM PC

FIGURE 3-3 Sample drawing produced on a CAD system. Courtesy of Versa CAD.

documenting the design process. It follows, then, that computer aided design and drafting, or CADD, is using a computer system to enhance the design and drafting processes.

The computer is particularly effective in releasing designers, engineers, and drafters from some of the more tedious and repetitive tasks which design and drafting involve. It gives them more time to devote to productive tasks. The computer is a particularly valuable tool in enhancing such typical design tasks as analysis, review, modeling, and testing through simulation. The computer is also a valuable tool for enhancing both 2-D and 3-D drafting.

Two-dimensional, or 2-D, drafting involves producing orthographic representations of design components. In an orthographic view of a subject, the viewer's line of sight is perpendicular to the object. There are six principal orthographic views of any object: top, front, right side, bottom, back, and left side (Figure 3-4). A fundamental rule of drafting is to construct as many views as are necessary to completely define the subject for production personnel, but only those views. Usually the top, front, and right-side views are used in 2-D drafting and are all that are necessary. However, on occasion, owing to the shape of an object or some other peculiarity, more than three views (and views other than the top, front, and right side) must be used.

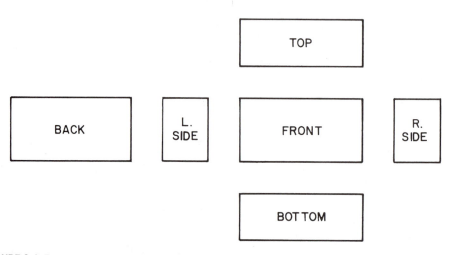

FIGURE 3-4 The six principal orthographic views of any object.

Because of the inherent capabilities of a computer, three-dimensional, or 3-D, drafting is becoming more and more common in CAD/CAM settings. A three-dimensional drawing more closely resembles how the human eye actually sees a subject than does a two-dimensional drawing. Consequently, it takes less skill on the part of the reader to interpret a 3-D drawing. Three-dimensional drawings can be the result of isometric, diametric, trimetric, oblique, or perspective projection (Figure 3-5).

Oblique perspective is sometimes referred to as 2½-D rather than 3-D. In any case, a 3-D drawing is one which shows the length, width, and depth of the subject in one view. Three-dimensional representations of designs in a CADD system are either wireframe rep-

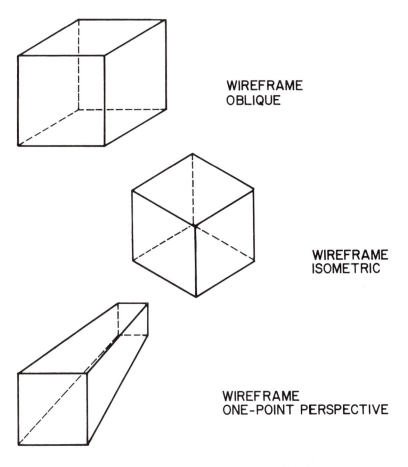

WIREFRAME
OBLIQUE

WIREFRAME
ISOMETRIC

WIREFRAME
ONE-POINT PERSPECTIVE

FIGURE 3-5 Types of three-dimensional projections.

resentations or solid models. Wireframe representations look exactly as the name implies; they can be seen through and appear to be constructed out of wire. Three-dimensional solid models are truer and more accurate representations of a design than wireframe models.

ADVANTAGES OF CADD

There are a variety of advantages over manual design and drafting which CADD offers designers, engineers, and drafters. Some of the more frequently stated advantages are:

1. improved productivity;
2. higher quality designs which lead to better products;
3. higher quality documentation packages;
4. elimination of tedious, repetitive tasks;
5. time savings;
6. opportunities for growth for design and drafting personnel;
7. better standardization;
8. common database;
9. improved interface between design and manufacturing;
10. decreased response time for bids.

Improved Productivity

The productivity improvement figures which can be expected as a result of a conversion from manual design and drafting to CADD will vary from company to company and from application to application. This is partially due to the fact that there are no standards available for calculating productivity improvements relative to CADD and partially because there are so many variables which affect the degree of improvement.

Such factors as the amount of training provided design, drafting, and engineering personnel; the degree to which design, drafting, and engineering personnel accept CADD; management's attitude toward CADD; and the creativity of individual users will all have an effect on the productivity gains that are actually realized in a given setting. Industry-wide averages for productivity improvement tend to range between 2:1 and 4:1 after a conversion to CADD.

Higher Quality Designs

Because CADD relieves designers of many of the tedious and repetitive tasks traditionally associated with design and drafting, the designers are able to devote more time to actually improving the design. Designers in a CADD setting tend to be less reluctant to make changes that will improve the design, because the changes will require less work than they would in a traditional manual setting. This is human nature, and it tends to lead to higher quality designs which, in turn, lead to better products.

Higher Quality Documentation Packages

CADD gives users the ability to produce neater, more consistent documentation. In addition, because it is so easy to produce pictorial drawings, rotated views, exploded views, and zoomed-in views, these things are more likely to be included in a CADD produced documentation package. Such additions to the documentation package make the job of manufacturing personnel easier.

Elimination of Repetitive Tasks

In the past, much of the time spent in the design process has been on nonproductive repetitive tasks which are tedious. A classic example of a tedious repetitive task intrinsic to traditional design and drafting is freehand engineering lettering. With a CADD system, lettering is accomplished by pressing buttons which, of course, takes significantly less time. Another example is dimensioning. Automatic dimensioning relieves CADD system users of the tedium of continually constructing extension lines, dimension lines, and arrowheads.

Time Savings

CADD offers a number of time-saving features. Such features as typing of text, symbols menus, automatic updating of revised dimensions, and such data manipulation capabilities as *rotate*, *mirror*, *copy*, and *move* save a significant amount of time in the design and drafting process.

Personnel Growth

After a long period of time on the job, many experienced designers, drafters, and engineers find they have reached the limits of their growth potential. They have little or nothing left to learn. When this is the case, work tends to become uninteresting. CADD opens up new avenues of learning and affords people the potential for continued growth. Even after one becomes proficient in the operation of a CADD system, there are continual new avenues to explore and new horizons to aspire to.

Standardization

In order for the CAD/CAM interface to work, design and manufacturing personnel must be speaking the same language. This forces both design and manufacturing personnel to learn standards and to stick to them. This has historically been a problem between design and manufacturing.

Common Database

The information developed during the design process is valuable information. It can be used by design, sales and marketing, manufacturing, accounting, contracts and bids, scheduling and production control, quality control, and shipping personnel. Through intercompany networking, the database created during the design process can be shared by all of these personnel.

The Design and Manufacturing Interface

The interface between design and manufacturing personnel has always been a problem. The definition of a good drawing is a drawing which contains all of the information needed by manufacturing personnel to produce the product without asking for clarification from design personnel. In a traditional manual design setting, the interface between design and manufacturing personnel has not always been a good interface. However, with CADD, the tendency toward improved standardization and the common database improves the design/manufacturing interface.

Faster Response to Bids

Manufacturing firms stay alive by winning enough bids to keep their companies profitable. Bids must be submitted within a specified time frame, and this time frame often is difficult to comply with. Seldom will bid personnel have all of the time they would like to prepare a comprehensive, well-planned, well-presented package. CADD, because it saves time in so many different ways, helps to solve this problem. Because of the advantages of CADD, manufacturing companies can respond faster to bids and with more attractive bid packages. This, in turn, allows them to bid on more jobs.

These are some of the more frequently stated advantages of CADD over traditional manual design and drafting techniques. You will notice that they tend to intertwine and overlap. That is, one advantage tends to be the product of another advantage. In spite of these advantages, CADD is not without its detractors. You should be familiar with some of the more common complaints about CADD.

COMPLAINTS ABOUT CADD

There are a number of concerns frequently expressed about CADD by people in design, drafting, and engineering.

1. CADD represents change, and even the most positive change is liable to be resisted by some personnel.
2. Because CADD is a relatively new concept, well-trained, highly skilled CADD technicians are hard to find and keep. The competition for these people is intense.
3. It may be difficult to justify the investment in CADD unless the system can be kept in operation continually for at least one eight-hour shift a day, and preferably two.
4. Highly complex top-end systems are very complicated and take a long time to learn to use effectively.
5. Because of the learning that must take place before a traditional designer will be productive on a CADD system, productivity ratios tend to decrease before they increase.
6. Management sometimes expects instant results or more than CADD is able to produce.
7. Initial costs for top-end systems can be extremely high.

APPLICATIONS OF COMPUTERS AND SOFTWARE IN DESIGN AND DRAFTING

There are a variety of applications of computers and software in design and drafting. The applications can be divided into two broad categories: CADD applications and non-CADD applications. CADD applications of computers include 2-D drafting, 2½-D drafting, 3-D drafting, and such design support functions as solid modeling, analysis, review and evaluation, and simulation. Non-CADD applications of computers include decision support, office automation, project management, and database management.

CADD Applications of Computers

One of the most important types of documentation produced on CADD systems is the engineering drawing or working drawing. Working drawings are produced in 2-D, 2½-D, and 3-D formats. Two-dimensional drawings are flat orthographic representations of the object; 2½-D drawings are oblique representations of the object; and 3-D drawings are isometric, diametric, trimetric, or perspective representations of an object. Three-dimensional drawings should not be confused with three-dimensional solid models, which will be discussed later in this section. Three-dimensional drawings on a CADD system are sometimes referred to as wireframe 3-D drawings to distinguish them from three-dimensional solid models.

Figure 3-6 is a drawing which contains both a 2-D and a 3-D representation of a wrench. The 2-D representation is the view on which sections "A" and "B" have been cut. The 3-D view is the one immediately above it. To the designer, drafter, and engineer, a 2-D, 2½-D, or 3-D drawing is a series of points, lines, and planes. However, to the computer, such drawings are mathematical representations of geometric models. A 2-D drawing to a computer is a series of X and Y coordinates. A 3-D drawing is a series of X, Y, and Z coordinates.

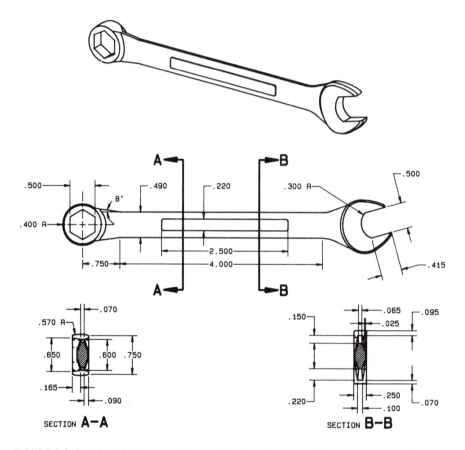

FIGURE 3-6 Drawing with two- and three-dimensional representations of a wrench. Courtesy of Tektronix, Inc.

Designers and engineers use computers for constructing solid models, analyzing designs, reviewing and evaluating designs, and testing designs through simulation. Designers use solid models to achieve a realistic representation of the part or object being designed. A solid model gives the designers a more accurate, more realistic representation of the design than does a 3-D wireframe drawing. A wireframe model describes only the exterior envelope of the object. A 3-D solid model describes both the interior and the exterior, and much more realistically. Designers use computers as tools in design analysis also.

All designs have certain analysis tasks inherently required in them. Such analysis tasks as heat transfer, stress, strain, and friction calculations may be accomplished rapidly and accurately using a computer. One of the most popular analysis features of a computer is its ability to conduct a finite element analysis of a geometric model (Figure 3-7).

FIGURE 3-7 Three-dimensional model for finite element analysis. Courtesy of McDonnell Douglas Automation Company.

In conducting a finite element analysis, the computer divides the geometric model into a number of finite elements, which are small rectangular or triangular shapes. Then, by analyzing each finite element, the computer is able to determine how it will behave under certain conditions, how each element will relate to the other elements when exposed to these conditions, and how the entire design will behave as a whole under these conditions. The computer is also a valuable tool in reviewing and evaluating designs.

Such computer capabilities as automatic dimensioning, automatic follow-through on dimension revisions, the zoom-in function, layering, and interface checking are invaluable in reviewing and evaluating designs. Automatic dimensioning cuts down on initial dimensioning errors. The follow-through capability ensures that when one dimension is changed on a design, all other affected dimensions are changed correspondingly. The zoom-in function allows designers to zoom in on intricate details and magnify them so they can be more easily seen and dealt with. Layering allows a design to be superimposed on patterns, templates, or drawings of rough castings. Interference checking allows designers to see graphically if one part that is supposed to fit into another part will actually do so in the finished product.

Simulation is a computer capability which allows designers to test certain functions of a design without having to create a working prototype or live model. Sometimes referred to as kinematics, this function allows designers to build a geometric model of the design, display it on the monitor, and introduce motion to see that it will actually work as it was designed to work.

Non-CADD Applications of Computers

The computer can be used for decision support functions such as accomplishing calculations, storing statistics, financial modeling, reporting, projecting, projection modeling, and a variety of other types of decision support. Office automation uses include word processing, electronic mail, electronic scheduling, and creating electronic spreadsheets. Project management applications consist of establishing specific tasks, time frames for those tasks, and tracking of progress in keeping to the time frame.

One of the main advantages of CADD is that it allows for a common database among design, manufacturing, quality control, sales and marketing, accounting, and a variety of other departments. The computer allows this common database to be interacted with and managed by design personnel and instantly shared with these other functional areas.

COMPUTER-AIDED DESIGN AND DRAFTING HARDWARE

Hardware is the collection of mechanical equipment and devices you see when you look at a CADD system. The nucleus of a CADD system is the processor, which is the computer in a computer-aided design and drafting system. Chapters 1 and 2 were devoted to a study of computers. Consequently, nothing further will be said about them in this section. Rather, this section will focus on the other hardware components or peripherals in a typical CADD hardware configuration: display terminals, input/interaction devices, and output devices.

Display Terminals

All CADD systems have display terminals. Some have one terminal which is capable of displaying both text and graphics. Other configurations use dual terminals, one for displaying graphics and the other exclusively for text. Figure 3-8 is an example of a single display monitor. Figure 3-9 is an example of a dual screen configuration.

FIGURE 3-8 Single display monitor option. Courtesy of CalComp, a Sanders Company.

FIGURE 3-9 Dual display monitor option. Courtesy of CalComp, a Sanders Company.

There are three types of display terminals used in CADD systems (Figure 3-10):

1. Raster displays
2. Refresh graphics displays
3. Storage tube displays

Raster Displays

Raster displays are the most frequently used types of terminals in CADD systems. In a raster display, an electron beam gun creates the image by illuminating the picture elements, or pixels, on the display screen. A beam from the electron gun scans the screen horizontally back and forth, illuminating the appropriate pixels.

An important concept with regard to raster displays is "resolution." Resolution is a term used to describe the quality of the image being displayed. A high-resolution display is capable of high quality images, while a low-resolution display will produce low quality images. A high-resolution image is smooth, crisp, and sharp. A low-resolution image is jagged, uneven, and rough. Whether a given display is a high-resolution

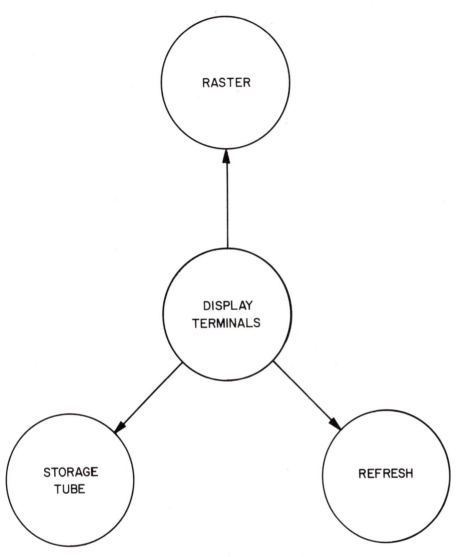

FIGURE 3-10 Three types of display terminals.

or low-resolution display is a function of the number of pixels available on the screen to be illuminated.

The more pixels available, the higher the resolution. A low-end display might have a screen with 260 X 380 pixels. Any display that has over 500 X 500 pixels can be considered a high-resolution display. However, there are top-end displays available with over 1,000 X 1,000 pixels.

Some raster displays have a capability known as *bit mapping*. Bit mapping allows individual pixels to be turned on or off by the computer. Bit mapping can decrease the amount of time required to create an image by raster scanning. However, a bit-map display requires considerably more memory than one without this capability. This is because each pixel on the screen that can be turned on or off must have one bit of memory available in the main memory.

Refresh Graphics Displays

A refresh graphics display creates an image by directing an electron beam from an electron gun against the back of the display screen. The electron beam draws the image in the phosphor coating of the screen. An image created in this manner has a very short life and will begin to flicker unless it is continually retraced by the electron beam. This process of retracing the image is known as "refreshing." Images on such a display must be refreshed from thirty to sixty times per second in order to maintain their quality.

Storage Tube Displays

A storage tube display is similar to a refresh display in the way the image is created. As in a refresh display, the image is created by an electron beam which draws a picture on the back of the display screen. However, unlike a refresh display, the image on a storage tube display does not have to be constantly refreshed. Rather, once created, it is stored. Storage tube displays can output data in two modes: temporary and permanent.

Data that is still being interacted with and might require changes or revisions is stored in the temporary mode. In this way, it can be easily and quickly altered. Once the data have been interacted with and have reached the point that no further changes are required, they are moved to the permanent mode, where they are more difficult to alter.

Input/Interaction Devices

When working with a CADD system, users need ways to get data into the system for processing and ways to manipulate those data once they are in the system. There are a variety of devices used for inputting and interacting with data in a CADD system. The most frequently used of these are the keyboard, the digitizer, the puck and the stylus, the joystick, the trackball, the mouse, and the light pen.

Keyboards

The keyboard is the most frequently used input and interaction device in most CADD systems. It can be used for entering commands, text data, and XY or XYZ coordinates when creating graphic data. Keyboards used in CADD systems contain the normal alphanumeric characters found on a typewriter keyboard, as well as a variety of special keys and, frequently, a separate numeric pad for speeding mathematical calculations. A typical keyboard in a CADD system, such as the one shown in Figure 3-11, is capable of 256 or more characters which are made using eighty to ninety keys singly or through multiple key interaction.

FIGURE 3-11 Keyboard with a separate numeric pad configuration. Courtesy of Auto-trol Technology Corporation.

Digitizers

Digitizers were originally developed in the early 1960s. This new technological development marked the critical initial linkage between a digital computer and the graphic requirements of design and drafting. The digitizer is the hardware device which encouraged the initial development of computer graphics software. A digitizer is an electromechanical device which is capable of converting graphic data into XY coordinates or digital data; hence, the name digitizer.

A digitizer is like a piece of electronic graph paper. It has a 0,0 origin point, X coordinates which progress along a horizontal axis, and Y coordinates which progress along a vertical axis. Graphic data being digitized are converted to positional data relative to the point of origin, then entered into the computer and stored as XY coordinates.

There are a variety of different types of digitizer technologies. There are mechanical, sonic, magnetostrictive, electrostatic, and electromagnetic digitizers. Regardless of the technology, all digitizers serve the same purpose. In addition to serving as a means for converting graphic data into digital data, digitizers can also be used as electronic drawing tablets, devices for controlling the screen cursor in conjunction with a puck or light pen, and as devices for mounting overlays for command and symbol menus. Figure 3-12 is an example of the type of digitizer frequently used in modern CADD systems.

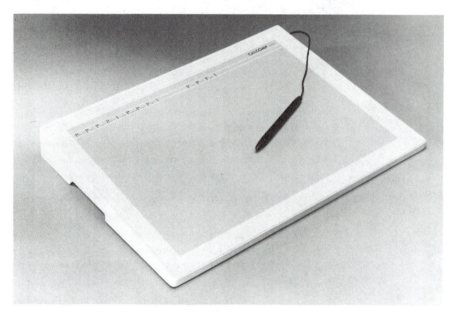

FIGURE 3-12 Series 2500 digitizer. Courtesy of CalComp, a Sanders Company.

As it is with display terminals, resolution is an important factor with regard to digitizers. Low-resolution digitizers are only capable of inputting low-quality images. High-resolution digitizers are capable of high-quality images. The resolution of a digitizer is a function of the number of X and Y coordinates it is capable of plotting. Digitizers with resolution of 0.001 of an inch or greater are generally required in modern CAD/CAM settings.

Puck and Stylus

The puck and the stylus are different forms of the same device. Both are electronic devices used to move the cursor across the display by moving it across the surface of a digitizing tablet. Figure 3-13 is an example of a stylus used in a CADD system.

FIGURE 3-13 Type of stylus used in some modern CADD systems. Courtesy of CalComp, a Sanders Company.

The Joystick, the Trackball, and the Mouse

The puck and stylus are used in conjunction with a digitizing tablet in controlling the screen cursor. The joystick, the trackball, and the mouse do not require a digitizer, but, like the puck and stylus, they are used to control the screen cursor. The joystick consists of a spring-loaded stick mounted in a small square or rectangular console. The stick can be tilted in any direction. The direction of the tilt guides the movement of the screen cursor correspondingly.

The trackball resembles any normal type of ball mounted in a small square or rectangular console. The trackball is rolled using the palm of the hand. The direction of movement controls the direction of movement of the screen cursor, and the speed of movement controls the speed of movement of the screen cursor.

The mouse is a small rectangular device on rollers. It operates like an upside-down trackball. As it is rolled across a smooth surface, the screen cursor moves correspondingly, matching the speed of movement of the mouse. The joystick and trackball are used less and less in modern hardware configurations. However, the mouse is seeing an increase in use. Figure 3-14 is an example of a mouse used with a modern CADD system.

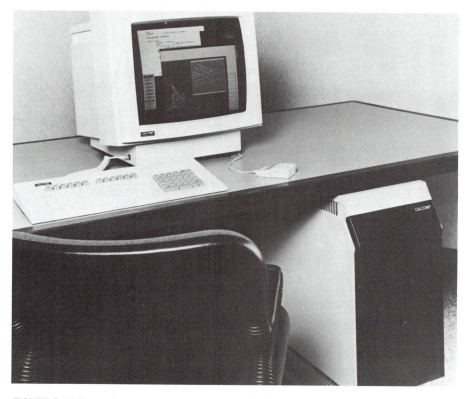

FIGURE 3-14 **Type of mouse used in some modern CADD systems.** Courtesy of CalComp, a Sanders Company.

The Light Pen

The light pen is a stylus-looking device which is used for interacting directly with the screen of the display terminal. Light pens detect the light which emanates from the display screen. Light pens are most commonly used for selecting menu options which are displayed on the terminal screen.

Output Devices

There are a variety of output devices that can be used in CADD systems; these include pen plotters, electrostatic plotters, computer output film devices, printer plotters, laserplotters, color ink-jet plotters, color copiers, CRT plotters, and photoplotters. The most frequently used of these are the pen plotter, electrostatic plotter, photoplotter, and printer plotter. These devices are used for converting digital data stored in the computer back into human-usable forms such as graphic and alphanumeric data. When examining output devices, it is important to consider their resolution, accuracy, and repeatability capabilities.

Resolution with a plotter means the same thing that it means with the display terminal and the digitizer. A high-resolution plotter is able to plot smooth, crisp, sharp lines. Plotters with at least one-thousandth of an inch resolution are necessary in modern CADD situations.

Accuracy is the same thing in CADD that it was in manual drafting. If a line is supposed to be six inches long, how close is it to actually being six inches long when plotted?

Repeatability is the capability of a plotter to repeat a line or a point and to position it exactly where it was drawn the first time. A high-repeatability plotter could draw a line and then draw back over the line without widening it at all.

Pen Plotters

Pen plotters are electromechanical devices which use especially designed pens in creating graphic and alphanumeric data on a variety of different types of media. Some pen plotters are flatbed in configuration and some are drum. On flatbed plotters, the medium is stationary while the pen moves. On drum plotters, both the medium and the pen move. Modern pen plotters come in single and multiple pen configurations. Figure 3-15 is an example of a flatbed plotter. Figure 3-16 is an example of a drum plotter.

Electrostatic plotters are a frequently used type of output device in CADD systems. However, since they are not able to produce the same quality of plot as a pen plotter, they are not generally used for finished documentation packages. An electrostatic plot resembles an old-time xerographic copy. Electrostatic plotters copy by applying charges to special paper that then moves through a toner bath. Special toner adheres itself to the charged spots and is cooked on in a subsequent step.

Electrostatic plotters are extremely fast when compared to pen plotters. For this reason, they are sometimes used for making hard copies of preliminary documentation and documentation that is going to remain in-house. Electrostatic plotters can also double as printers. Figure 3-17 is an example of a monochrome electrostatic plotter. Figure 3-18 is an example of a color-capable electrostatic plotter.

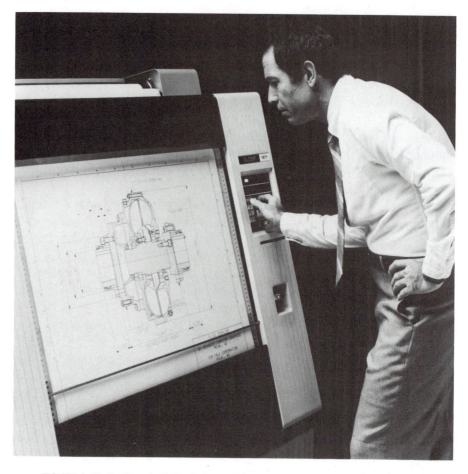

FIGURE 3-15 Dual-mode flatbed plotter. Courtesy of CalComp, a Sanders Company.

Photoplotters are mechanically structured like pen plotters. However, rather than creating images using special pens, they create images using a light beam that inscribes the image on photographic film which is then developed in a darkroom. Photoplotters are extremely expensive and are not used except in those applications requiring maximum levels of accuracy. The most common applications of photoplotters are in CADD systems used in the design of printed circuit boards and integrated circuits.

Printer Plotters

Some types of printers can double as plotters. The most commonly used type of printer plotters in CADD systems are impact, thermal, and electrophotographic printer plotters.

Impact printer plotters create an image when some type of device, such as a key or a daisy wheel, strikes an ink-bearing ribbon

FIGURE 3-16 1040GT drum plotter. Courtesy of CalComp, a Sanders Company.

against the paper in a manner similar to a typewriter. Impact printer plotters create graphic images using a series of tiny dots.

Thermal printer plotters use a special heat process that develops an image on specially treated paper.

Electrophotographic printer plotters use a special dry, silver-coated, photosensitive paper, and develop the image through a special heat process.

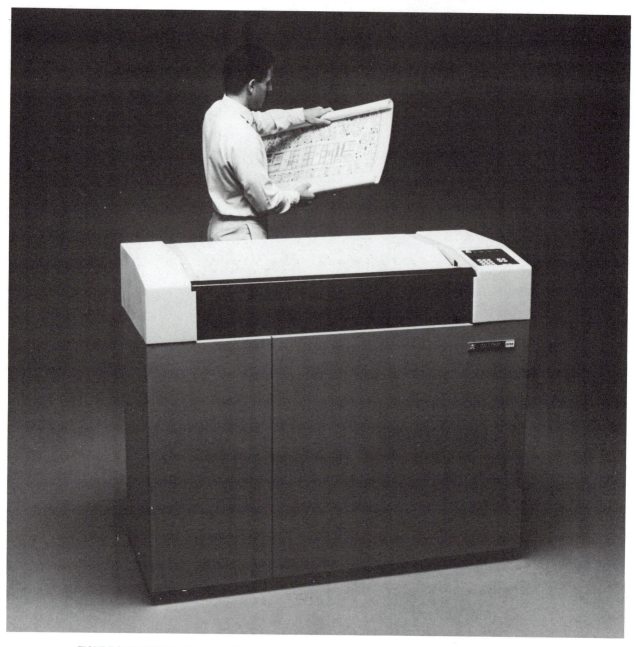

FIGURE 3-17 5700 Series monochrome electrostatic plotter. Courtesy of CalComp, a Sanders Company.

Thermal and electrophotographic printer plotters are faster than pen plotters, but they sacrifice quality for speed. The image produced by thermal and electrophotographic printer plotters is a poor-quality image. Figure 3-19 is an example of a printer plotter used in a modern CADD system.

FIGURE 3-18 5800 Series color electrostatic plotter. Courtesy of CalComp, a Sanders Company.

CADD SYSTEM OPERATION

CADD software is written to allow for ease of interaction between humans and the computer system. There are a wide variety of graphics packages available for the various CADD systems on the market.

In some ways, these graphic packages vary drastically. However, they are very similar in other ways. The various categories of functions available to users of CADD systems are (1) graphic data creation, (2) graphic data modification, (3) graphic data manipulation, (4) display control

FIGURE 3-19 ColorMaster plotter/printer. Courtesy of CalComp, a Sanders Company.

and manipulation, (5) graphic data specification, (6) graphic data facilitation, and (7) graphic data output.

Under each of these major headings, each CADD system will have a variety of individual commands. These commands differ from system to system, as do the methods used for entering the commands. Some CADD systems are command-driven via commands entered through the keyboard. Some systems are menu-driven via commands entered from screen displayed menus, menus mounted on digitizer tablets, or special keyboard menus. In spite of their individual differences, all CADD systems will have at least these major categories of functions.

Graphic Data Creation

Graphic data creation functions are the functions designers and drafters use to actually create the graphic image. Graphic images are composed of points, lines, planes, and a variety of geometric characters such as circles, squares, rectangles, ellipses, triangles, and so on. The types of commands typically grouped in this category are:

points	ellipse
solid lines	polygon
hidden lines	grid
construction lines	XYPT
dash lines	polar
center lines	line weight
rectangle	fillet
circle	chamfer
arc	leaders
irregular curve	hatch

Using commands such as these and others which fall into this broad category, designers and drafters are able to create any type of graphic image required to document the design process.

Data Modification Commands

When creating graphic data, it is frequently necessary to modify, or edit, what has been created. Data modification commands allow designers and drafters to do this. Typical commands grouped in this category include EDIT, DELETE, REDRAW, CLIP, and ERASE.

Commands which fall into this category give designers and drafters great latitude in making corrections and revisions to designs and the documentation of those designs.

Manipulation Commands

Manipulation commands allow designers and drafters to make transformations, to reposition, or to change in some way the graphic image as it is displayed on the monitor. Commands grouped in this functional category include MOVE, COPY, ROTATE, SCALE, and MIRROR.

The MOVE command allows users to reposition the graphic image or some portion of it. The COPY command allows users to quickly copy exact replicas of a graphic image which has already been created once. The ROTATE command allows users to revolve a graphic image around an XY or XYZ axis. The SCALE function allows users to actually increase or decrease the size of a graphic image. The MIRROR function allows users to create a mirror image of a graphic image they have already created. The original image may be retained or eliminated.

Display Control and Manipulation

In addition to manipulating data, designers and drafters are able to manipulate the screen of the display monitor. The display manipulation function allows users to view the graphic image from different perspectives and different distances. Typical commands grouped in this functional category include ZOOM and PAN.

The ZOOM command, in essence, moves the viewer's eye closer to or farther away from the graphic image, causing the image to appear to magnify or reduce with each movement. The PAN command allows users to shift the screen around to view different sections of a zoomed-in on object while still in the zoom mode.

Size Specification Commands

When creating graphic data to document the design process, it is usually necessary to specify size and other characteristics of the data. Size specification commands allow users to do this. Commands typically grouped in this functional category include AUTOMATIC DIMENSION, DIMENSION, and TOLERANCE.

These commands allow designers and drafters to do what has traditionally been called dimensioning.

Facilitation Commands

Beyond the need to create, modify, manipulate, and specify data, there is also the need to facilitate data when using a CADD system. Facilitation commands do not create or alter data. Rather, they allow the data to be saved and allow users to interact with the system to save or call up data. Typical commands found in this category include SAVE, FILE, QUIT, and DEFAULT.

Output Commands

Output commands are an important part of any graphics package in that they allow designers and drafters to get graphic images and any other type of data they have created out of the CADD system in human-usable form. Most of the time this means that the data will be either plotted or printed. Consequently, typical output commands are simply *plot* and *print*.

CAD/CAM INTEGRATION AND NETWORKING

In a manufacturing firm there are several different departments which produce data. These departments include design and drafting, accounting, purchasing, scheduling, contracts and bids, production, quality control, and several others. Each department produces data that are helpful to the others.

Traditionally these various departments have communicated through various forms of hard copy (i.e., drawings, computer printouts, reports, specifications, etc.). With the advent of CAD/CAM, people began to work toward a new approach to communication: electronic communication directly from system to system via local area networks or LANs. A LAN is a communication pathway which allows for the transfer of data from one computer system to another.

LANs are already in use in design departments which have traditional minicomputer-based CADD systems. The network connects the system's various workstations so that a designer at one terminal can transfer data directly to a designer at another workstation. At present, this is possible when all of the pertinent hardware within the system comes from the same vendor. When this is not the case, problems arise.

It is this "compatibility" problem that is currently holding back the complete networking of all the various departments within manufacturing firms. LANs are not currently able to facilitate direct system-to-system communication among computers, robots, CNC machining centers, and other computer-controlled systems that are produced by different vendors.

Because vendor compatibility is the major inhibitor of total integration through networking, several organizations are attempting to solve this problem by drafting standards which will lead to international compatibility in data communication. The leader among these organizations is ISO, or the International Standards Organization. A specification called IGES (International Graphic Exchange Specification) is a step toward standardization that will solve the networking problem.

Once the communication problems have been solved, a number of management problems will still remain. These include answering such questions as:

1. Who owns what data?
2. Who has access to what data?
3. How is access controlled?

These questions must be dealt with at the local level by managers in individual companies.

THE CADD MARKET

Since its inception in the late 1960s, CADD has become a multimillion dollar business. In the late 1960s, there were only a handful of CADD vendors. Now there are over 100 traditional vendors, and over 150 microCADD vendors. Traditional CADD systems are those based on minicomputers and mainframe computers. MicroCADD systems are those based on microcomputers.

Before the early 1980s, although there were several microCADD systems on the market, microCADD was not a realistic option for the private sector. Prior to this time, microcomputers did not have the processing speed, memory capacity, or graphic capabilities needed in computer-aided design and drafting situations.

However, with the development of the IBM personal computer, or PC, microCADD became a feasible option. This was the first microcomputer with the memory capacity, processing speed, and graphic capabilities needed in CADD. IBM has followed its original PC with the even more powerful AT and XT models. A number of other vendors have developed IBM "Work-a-like" microcomputers. Consequently, by 1985, microCADD represented a legitimate and rapidly growing portion of the overall CADD market.

Figure 3-20 is a graph of the approximate sales revenues for 1985 for the leading traditional CADD vendors. These vendors are:

IBM

Intergraph

Computervision

CALMA

Mentor

Daisy

McAuto (McDonnell Douglas Automation Company)

Control Data

Applicon

Together, these nine vendors represent over 76% of the total traditional CADD market. Traditional CADD systems are available in costs ranging from approximately $50,000 to well over $1,000,000. However, a typical traditional CADD system will cost between $50,000 and $250,000. It was costs such as these that eventually led to the development of less expensive microCADD systems.

MicroCADD is the fastest growing portion of the overall CADD market. While the annual revenue growth rate for traditional CADD systems is in the 35% to 38% range, the annual growth of revenue in microCADD is over 60%. Because microCADD systems usually range in price from $10,000 to $25,000, they will probably never approach the level of revenue generated by traditional systems. In 1984, the total revenue generated by the sales of CADD systems, services, and products was over $3 billion. Of this, only $67 million can be attributed to microCADD. However, the number of installed systems based on microcomputers is growing rapidly.

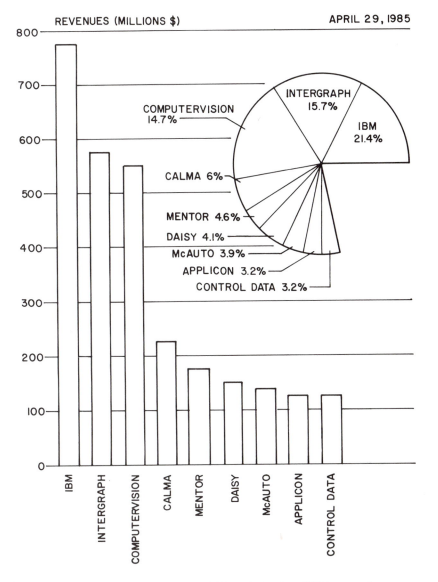

FIGURE 3-20 Sales revenues for 1985 for the leading CADD companies.

So rapidly, in fact, that by 1990 there will be more microCADD systems in use than traditional CADD systems. Figure 3-21 contains a chart showing the market shares of the leading vendors of microCADD systems. The leaders in this market are:

Autodesk   Daisy
Chessel-Robocom   Futurenet
P-CADD   Cascade
VersaCAD

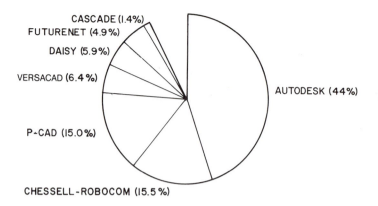

FIGURE 3-21 Market shares of leading microCADD vendors.

Notice that Autodesk has an even larger share of the microCADD market than IBM has of the traditional CADD market. Figure 3-22 contains a graph showing the rapid growth in sales experienced by Autodesk during 1984. As microcomputers continue to improve in memory capacity, processing speed, and graphic capabilities, microCADD will become the norm in computer-aided design and drafting.

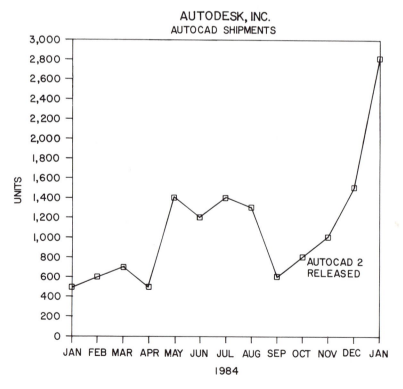

FIGURE 3-22 Growth of the leading microCADD vendor in 1984.

CADD-RELATED MATH

Designers and drafters are frequently required to use math in accomplishing such tasks as thread, gear, speed and feed, and taper calculations. In order to perform such calculations, students must be proficient in the use of a number of standard algebraic formulas. This section will review some of the more frequently used formulas in the design setting and the ways in which they are applied.

Thread Calculations

Figure 3-23 illustrates the terminology used in thread calculations and the symbols for the terms. The formulas themselves are summarized below.

TO FIND	AMERICAN STANDARD FORMULA	METRIC FORMULA
Pitch	$p = 1/n$	$p = 1/n$
Numbers of threads per inch or millimeter	$n = 1/p$	$n = 1/p$
Flat at crest, external thread	$f_1 = 0.125 \times p$	$f_1 = 0.125 \times p$
Flat at root, internal thread	$f_2 = 0.125 \times P$	$f_2 = 0.125 \times p$
Flat at crest, internal thread	$f_3 = 0.250 \times p$	$f_3 = 0.250 \times p$
Depth, external thread	$d_1 = 0.61343 \times p$	$d_1 = 0.61343 \times p$
Depth, internal thread	$d_2 = 0.54127 \times p$	$d_2 = 0.54127 \times p$

Gear Calculations

Figure 3-24 illustrates the various terms used in gear calculation formulas. Figure 3-25 summarizes the symbols for these terms and the formulas most frequently used in performing gear calculations.

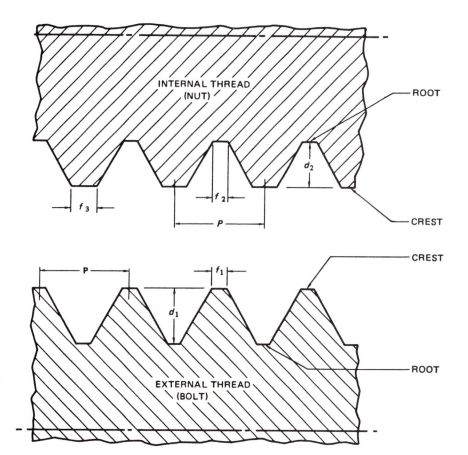

SYMBOL	MEANING
p	pitch
n	number of threads per inch or millimetre
f_1	flat at crest, external crest
f_2	flat at root, internal thread
f_3	flat at crest, internal crest
d_1	depth, external thread
d_2	depth, internal thread

FIGURE 3-23 Thread terminology and symbols. (From PRACTICAL PROBLEMS IN MATHEMATICS FOR MACHINISTS, 1980, Hoffman, Delmar Publishers Inc.)

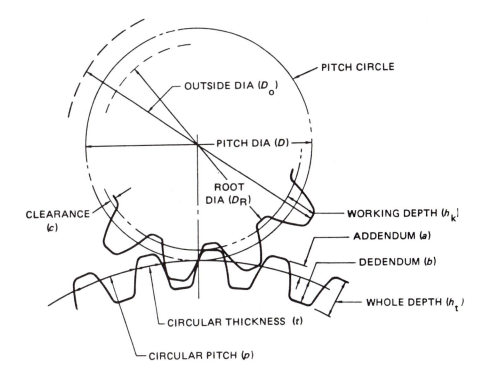

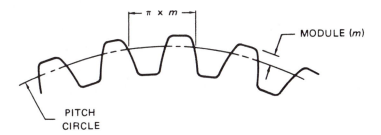

FIGURE 3-24 Gear terminology. (From PRACTICAL PROBLEMS IN MATHEMATICS FOR MACHIN-
ISTS, 1980, Hoffman, Delmar Publishers Inc.)

SYMBOL	MEANING
D	pitch diameter
D_o	outside diameter
D_R	root diameter
P	diametral pitch
p	circular pitch
t	circular thickness

SYMBOL	MEANING
a	addendum
b	dedendum
c	clearance
h_t	whole depth
h_k	working depth
n	number of teeth
m	module

TO FIND	AMERICAN NATIONAL STANDARD	METRIC
Module	· · · · · · · · · · · · · · · · ·	$m = \dfrac{D}{n}$
Diametral pitch	$P = \dfrac{n}{D}$	$p = \dfrac{1}{m}$
Pitch diameter	$D = \dfrac{n}{p}$ $D = \dfrac{D_o \times n}{(n+2)}$	$D = n \times m$
Number of teeth (expressed as a whole number)	$n = P \times D$	$n = \dfrac{D}{m}$
Addendum	$a = \dfrac{1}{P}$	$a = m$
Dedendum (preferred)	$b = \dfrac{1.250}{P}$	$b = 1.250 \times m$
Clearance (preferred)	$c = \dfrac{0.250}{P}$	$c = 0.250 \times m$
Clearance (minimum)	$c = \dfrac{0.157}{P}$	$c = 0.157 \times m$
Circular thickness-Basic	$t = \dfrac{1.5708}{P}$	$t = 1.570\ 8 \times m$
Root diameter	$D_R = \dfrac{(n - 2.5)}{P}$ $D_R = D - (2 \times b)$	$D_R = D - (2.5 \times m)$
Outside diameter	$D_o = \dfrac{(n + 2)}{P}$ $D_o = D + (2 \times a)$	$D_o = m \times (n + 2)$ $D_o = D + (2 \times m)$
Whole depth (preferred)	$h_t = \dfrac{2.250}{P}$ $h_t = a + b$	$h_t = a + b$
Working depth	$h_k = \dfrac{2}{P}$ $h_k = a + b - c$	$h_k = 2 \times a$
Circular pitch	$p = \dfrac{3.1416}{P}$	$p = 3.141\ 6 \times m$

FIGURE 3-25 Gear calculation symbols and formulas. (From PRACTICAL PROBLEMS IN MATHE-MATICS FOR MACHINISTS, 1980, Hoffman, Delmar Publishers Inc.)

Speed and Feed Calculations

Figure 3-26 illustrates the various symbols for speed and feed terms and the formulas most frequently used in performing speed and feed calculations.

SYMBOL	MEANING	AMERICAN STANDARD UNITS	METRIC UNITS
V	cutting speed	feet per minute (fpm)	metres per minute (m/min)
D	diameter	inches (in.)	millimetres (mm)
N	spindle speed	revolutions per minute (rpm)	revolutions per minute (r/min)

TO FIND	AMERICAN STANDARD UNITS	METRIC UNITS
N	$N = \dfrac{12 \times V}{3.1416 \times D}$	$N = \dfrac{1\,000 \times V}{3.141\,6 \times D}$
V	$V = \dfrac{3.1416 \times D \times N}{12}$	$V = \dfrac{3.141\,6 \times D \times N}{1\,000}$
D	$D = \dfrac{V \times 12}{3.1416 \times N}$	$D = \dfrac{V \times 1\,000}{3.141\,6 \times N}$

FIGURE 3-26 Feed and speed calculation symbols and formulas. (From PRACTICAL PROBLEMS IN MATHEMATICS FOR MACHINISTS, 1980, Hoffman, Delmar Publishers Inc.)

Taper Calculations

Figure 3-27 illustrates the various symbols for taper terms and the formulas most frequently used in performing taper calculations.

SYMBOL	MEANING
tpi	taper per inch
tpf	taper per foot
D	diameter, larger end
d	diameter, small end
L	length (in inches)

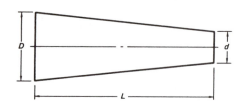

TO FIND	KNOWN	FORMULA
tpf	tpi	$tpf = tpi \times 12$
tpf	D, d, L	$tpf = 12\left(\dfrac{D-d}{L}\right)$
tpi	tpf	$tpi = \dfrac{tpf}{12}$
tpi	D, d, L	$tpi = \dfrac{D-d}{L}$
d	D, L, tpf	$d = D - \left[L\left(\dfrac{tpf}{12}\right)\right]$
D	d, L, tpf	$D = d + \left[L\left(\dfrac{tpf}{12}\right)\right]$
L	D, d, tpf	$L = 12\left(\dfrac{D-d}{tpf}\right)$

FIGURE 3-27 Taper calculation symbols and formulas. (From PRACTICAL PROBLEMS IN MATHE-MATICS FOR MACHINISTS, 1980, Hoffman, Delmar Publishers Inc.)

Chapter Three REVIEW

1. List and explain the five steps in the design process.
2. What is drafting?
3. What is Computer-Aided Design and Drafting (CADD)?
4. Explain the two categories of applications of computers and software in design and drafting.
5. List the three types of display terminals used in CADD systems.
6. List the most frequently used input/interaction devices.
7. How are the different CADD graphics packages similar?
8. What is a LAN?

Automation of machine tools began to see wide-scale implementation in 1975 with the advent of CNC machines. When manufacturing is fully integrated, CNC machines will represent one component of an overall CIM system.

Major Topics Covered

- What Is Numerical Control (NC)?
- History of NC
- CNC Machines
- CNC Input Media
- CNC Tape Formats and Code Standards
- Advantages and Disadvantages of CNC
- CNC Applications
- CNC Systems
- CNC Programming Coordinates
- Writing CNC Programs
- CNC-Related Math

Chapter Four

Computer Numerical Control

case study

Prosperity and even survival in the U.S. steel industry requires some sort of edge these days. Ellwood City Forge Corp. (ECF) of Ellwood City, PA, has found one, in the form of a knowledge-based scheduler.

A manufacturer of open die forgings, ECF handles orders that generally consist of no more than two or three parts, parts that are needed in small production lots or—due to size or strength requirements—can't be cast. The parts range in weight from 500 to 50,000 lb (225 to 22,500 kg). To do battle against tough offshore competitors, the company must be able to deliver parts quickly at an economical price. This goal is often at odds with its need to maximize manufacturing cost efficiency by batching orders for long production runs, minimal setups, full furnace loads, and high raw material yields.

The scheduling problem is twofold. The front-end problem involves the point of sales, where the company must be able to give a quote to a potential customer regarding the cost and time required to manufacture a particular part. Then, down on the shop floor, the new order must be folded into the existing production schedule for the three successive hot processes: melting, forging, and heat treating. Typically, there are some 2,000 open orders at ECF, each involving a unique sequence of operations.

Manual Scheduling

On the shop floor, manual scheduling is piecemeal in nature. Because it is too complex a job for one person, several schedulers share the responsibility, each trying to optimize the efficiency of a particular area. Unfortunately, this system does not allow optimization of the entire system— only parts of it.

One scheduler, in charge of the forging operations, back-plans to the New Castle (PA) plant where the specialty-grade steels are melted and cast into ingots. This scheduler must see that the ingot yield is as high as possible while maintaining a smooth, regular supply to the forging operations.

Courtesy of "Manufacturing Engineering", May 1987, "CIM Technology" Section. "Gaining the Upper Hand with AI" by Rita Schreiber.

Gaining the Upper Hand with AI

The job of the second scheduler, who takes over after the forging operations, is to schedule the various flows of parts for heat treatment.

Scheduling the heat treating area is particularly complex. There are a variety of furnaces suited for different processes. Oil, polymer, and water quenches are located adjacent to certain furnaces. Air-cooling procedures, required for certain parts, must also be scheduled. The issue of furnace fullness is a major consideration for gas efficiency, as is the issue of reserving particular furnaces for one or more parts of a batch as they return from quenching.

What the knowledge-based system will do is provide a single integrated procedure to balance the numerous and often conflicting objectives of the three hot operations. It will permit many constraints—far more than could be handled by manual schedulers—to be applied simultaneously. It will also allow various scheduling options to be evaluated quickly and accurately.

Says ECF President David Barensfeld, "The operator will be able to make iterations and reiterations of a proposed schedule very quickly. 'What-if' games can be played, taking the sting—and some of the cost—out of the warfare between fast delivery and optimal batching."

System Modules

The scheduling system consists of several modules: the knowledge base, automatic and interactive schedulers, interface to ECF's MRP-II database, and the simulation model.

The knowledge base is essentially a mini-model of the plant describing the available resources, orders, objectives, and constraints. This information is arranged in a schematic structure that allows factual knowledge to be separated from procedural knowledge.

The automatic scheduler makes decisions and propagates constraints and timeframes in which jobs must be done. The interactive portion

(continued)

of the scheduler allows the user to second-guess solutions, or to change the constraints and rerun the scheduler based on new information that isn't part of the scheduling logic yet.

The interface to the MRP-II database, which is under installation, is a mechanism to translate relevant data from the standard file structure of the MRP-II system into the scheduling system's knowledge-based structure. This translator passes on such information as process flow and due dates.

The simulation model has two modes. In one mode, the scheduler operates on its own, taking data from the plant and making suggestions for the next shift, day, or week. In the other, the user can initiate different scenarios himself, testing scheduling decisions with a variety of contingencies, such as unexpectedly long process times.

The new scheduling system, under development by Carnegie Group Inc. (Pittsburgh), is being installed in phases. The first phase, to be delivered this summer, will focus on improving the melt shop and forging operations. This phase will evaluate the tradeoffs between material yield, fuel costs, machine production, and customer service when orders for melting ingots are batched for forging. The next phase will focus on evaluating the tradeoffs between fuel costs, furnace utilization, and customer service when orders are batched for heat treatment. The final phase will address the integration of the melting, forging, and heat-treatment operations by using scheduling techniques to reduce manufacturing lead time.

"Traditional MRP systems assume a parts explosion of a predefined bill of materials for a single end product," says Ivan Johnson, director of contract systems development at Carnegie Group. "Or, from a material flow viewpoint, various purchased parts are assembled into a single end product. Conversely, in the steel industry, many end products with different shapes, sizes, and metallurgical properties are derived from a single melt of steel. Therefore, what we're dealing with here could be seen as a parts 'implosion.'

"More significantly," Johnson continues, "to produce these end products, similar but not identical intermediate products must be dynamically batched for production—for example, batching jobs into ingots. How well these batches are put together determines plant productivity, equipment utilization, and yield. To sift through the myriad of possible batches to find the best batch and then to schedule the batches requires much metallurgical, process, and scheduling knowledge. That's the technology gap that we're filling at ECF."

Machine tools such as lathes, mills, and drills form the nucleus of manufacturing shops which produce metal products. Historically, such tools were controlled manually. However, in the 1950s a new method was developed for controlling machine tools. That new method came to be known as numerical control, or NC.

WHAT IS NUMERICAL CONTROL (NC)?

Numerical control is a method of controlling machine tools using coded programs. These programs consist of numbers, letters, and special characters which define the path of the machine tool in accomplishing specific tasks. When the job changes, the program must be rewritten to accommodate the change. The programmability feature is the key to the growth of NC. Machine tools that are programmable are more flexible than their nonprogrammable counterparts.

NC is a broad term which encompasses the traditional approach to NC as well as the more modern outgrowths of computer numerical control (CNC) and direct numerical control (DNC).

HISTORY OF NC

Numerical control was first used in industry in the mid-1980s. However, its original development goes back to the late 1940s. The pioneer who began the development of NC was John Parsons. Parsons' company was involved in the development and production of aircraft products. Parsons was interested in NC as a way to improve his company's productivity in manufacturing these products.

With a contract from the Air Force, he began work on the development of the first NC machine. Parsons got things started, but he was not able to complete the project. After two years, the project was turned over to scientists and engineers at the Massachusetts Institute of Technology (MIT). In 1952, MIT completed the work Parsons had begun in 1949.

The first NC machines used punched cards and then punched tape as the programming medium. Punched tape was better than manual control, but it did have some disadvantages. Paper tape was easy to damage and difficult to correct or edit.

Mylar tape began to replace paper tape as a less fragile medium. This solved the problem of frequent damage, but it did not make punched tape any easier to edit. Such problems, coupled with advances in microelectronic technology, led to the use of computers as the means for programming machine tools. Computers solved the damage and editing problems. This gave birth to both CNC and DNC. CNC involves using a computer in writing, storing, and editing NC programs. DNC involves using a computer as the controller for one or more NC machines (Figure 4-1). A more advanced form of DNC is called "distributive numerical control" (Figure 4-2).

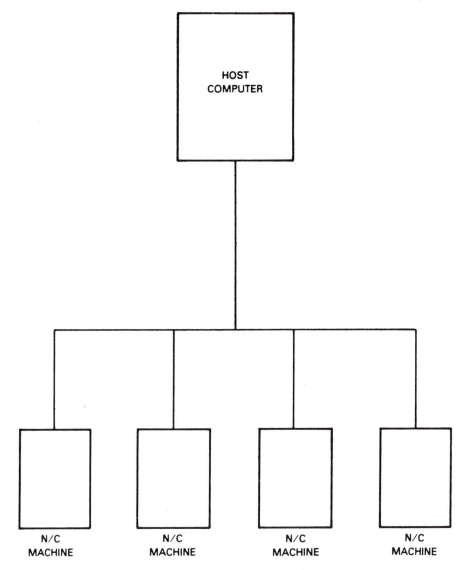

FIGURE 4-1 Direct numerical control. (From COMPUTER NUMERICAL CONTROL: CONCEPTS AND PROGRAMMING, 1986, Seames, Delmar Publishers Inc.)

In this concept, there is a host coupler and several intermediate computers which are tied to NC machines, robots, and other NC manufacturing machines. The main host computer is the central repository

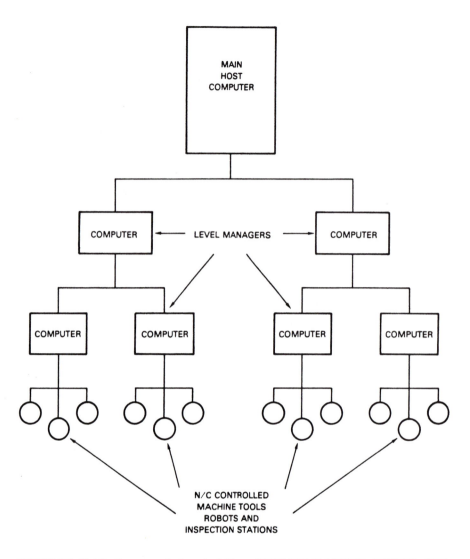

FIGURE 4-2 Distributive numerical control (From COMPUTER NUMERICAL CONTROL: CON-
CEPTS AND PROGRAMMING, 1986, Seames, Delmar Publishers Inc.)

for all programs for all jobs. Specific programs for specific jobs are
"dumped" from the host computer to the intermediate computers. This
cuts down on the amount of time required to get the programmed
instructions from the computer to the machine tool. Figures 4-3 and 4-4
are examples of the types of computers which can be used to program
CNC machines.

FIGURE 4-3 MacEZ-CAM II. Courtesy of Bridgeport Machines, Inc.

CNC MACHINES

Most manufacturing machines fall into one of four categories:

1. Cleaning and finishing
2. Inspection and quality control
3. Pressing and forming
4. Material removal

Cleaning and finishing machines perform such operations as coating, degreasing, deburring, washing, honing, lapping, buffing, and grinding. Cleaning and finishing machines are not strongly associated with numerical control. However, there are several CNC deburring, buffing, and grinding machines on the market.

The ACME Manufacturing Company produces a three-axis single-spindle machine for grinding, deburring, and buffing. The SurfTran Division of Robert Bosch Corporation produces a five-station CNC deburring machine designed to thermally debur small parts.

CNC machines also are available in the area of inspection and quality control. The FETTE Division of Scarberg Intertool produces the PKU inspection center, a CNC machine which inspects holes, shaver cutters, spiral bevel gears, and helical gears. The A.F. Green Company produces a CNC quality control system for production and research applications. Digital Electronic Automation produces a high-accuracy CNC

FIGURE 4-4 EZ-CAM II. Courtesy of Bridgeport Machines, Inc.

coordinate measuring machine. ITW Illitron produces a CNC gear-inspection machine which analyzes and measures gears, splines, and other piece parts.

Material removal machines are those most strongly associated with CNC. There are CNC mills, drills, lathes, saws, and grinders, as well as CNC machining centers, from a long list of vendors. Some of the leading vendors in this area are Bridgeport, Cincinnati Milacron, Republic-Lagun Machine Tool Company, Gullmeyer and Livingston Company, Enterprise Machine Tools, and Armstrong-Blum Manufacturing Company.

There are also CNC presswork and forming machines available. Strippit Di-Acro produces a CNC press brake. Schwarze-Wirtz of West Germany produces a CNC pipe-bending machine. Comeq, Inc., produces a CNC plate-bending machine.

These are only representative examples. There are other CNC machines and other vendors. However, these examples make the point

FIGURE 4-5 Series II CNC INTERACT 520V Bed-Type Vertical Machining Center. Courtesy of Bridgeport Machines, Inc.

that there are CNC machines available in all four of the major categories of manufacturing machines.

It is well for students of CAD/CAM to be familiar with the types of CNC machines used in modern manufacturing plants. Figures 4-5 through 4-9 are some examples of modern CNC machines. When examining CNC machines, there are a number of factors to look for. These factors vary according to the type of machine, but some commonly asked questions about such machines are:

1. What size is it?
2. How much does it weigh?
3. What is the tooling range?

FIGURE 4-6 INTERACT 1 Series I CNC Iron. Courtesy of Bridgeport Machines, Inc.

4. What is the size of the working surface?
5. What is the maximum load?
6. What is the positioning speed?
7. What is the accuracy rate?
8. What is the repeatability rating?

There are other questions that might be asked and some that should be asked about specific CNC machines. For example, if the machine is a mill, it is important to know the throat distance and the distance from the table to the spindle. These questions are representative of the types of criteria applied in the examination of a CNC machine.

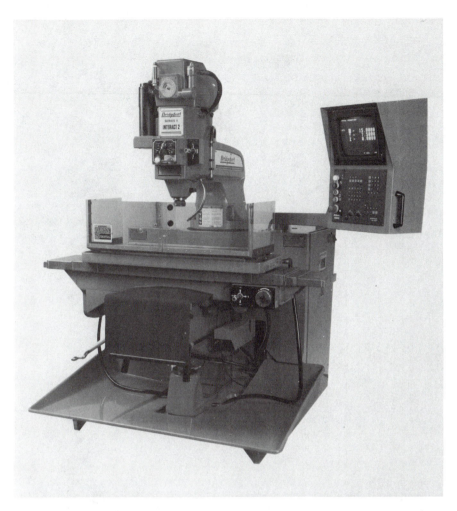

FIGURE 4-7 INTERACT 2 Series II CNC Iron. Courtesy of Bridgeport Machines, Inc.

Figure 4-5 is a Bridgeport Series II CNC Interact 520 bed-type vertical machining center. The Interact 520 is 45.8 inches by 19.7 inches and weighs 8,800 pounds. It has a range of 20 inches in the X axis, 15.7 inches in the Y axis, and 17.7 inches in the Z axis. The positioning speed is 590 inches per minute, and the repeatability is plus or minus 0.0002 inches. The accuracy ratings are plus and minus 0.00032 inches in the X and Y axes, and plus or minus 0.0004 inches in the Z axis. The working surface is 34 inches by 14.9 inches. The maximum load is 1,000 pounds.

Figure 4-6 is a Bridgeport Interact 1 CNC milling machine. The Interact 1 occupies 75 inches by 65 inches of floor area, is 87 inches high, and weighs 3,150 pounds. Its range is 18 inches in the X axis and 12 inches in the Y axis. The throat distance is 14.5 inches, and the table-to-spindle distance is 7.1 inches. The Interact 1 has a maximum load of 300

FIGURE 4-8 INTERACT 4 Series II CNC Iron. Courtesy of Bridgeport Machines, Inc.

pounds, a working surface of 34 inches by 12.5 inches, and a positioning speed of 197 inches per minute. The accuracy rating is plus or minus 0.0008 inches and the repeatability rating is plus or minus 0.0003 inches.

Figure 4-7 is a Bridgeport Interact 2 milling machine. The Interact 2 occupies 80 inches by 70 inches of floor area, is 89 inches high, and weighs 4,664 pounds. Its range is 29.9 inches in the X axis and 14.5 inches in the Y axis. The throat distance is 15.5 inches and the table-to-spindle distance is 7 inches. The Interact 2 has a maximum load of 1,000 pounds, a working surface of 38 inches by 15 inches, and a positioning speed of 200 inches per minute. The accuracy rating is plus or minus 0.0008 inches and the repeatability rating is plus or minus 0.0003 inches.

Figure 4-8 is a Bridgeport Interact 4 milling machine. The Interact 4 occupies 119 inches by 75 inches of floor area, is 90.75 inches high,

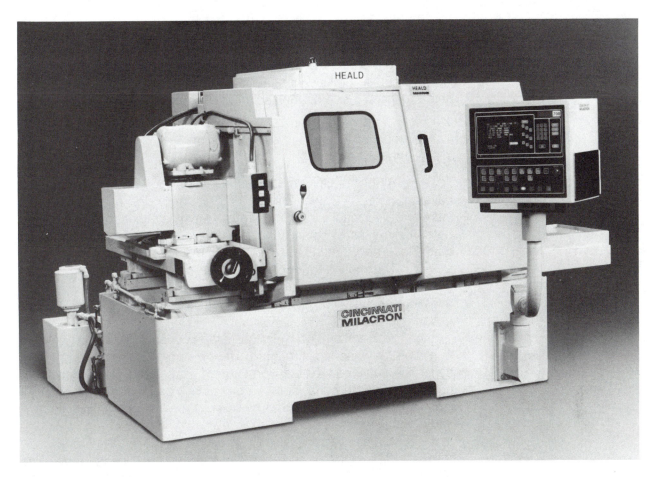

FIGURE 4-9 2EF CINTERNAL Multi-Surface Grinding Machine. Courtesy of Cincinnati Milacron.

and weighs 5,280 pounds. Its range is 29.9 inches in the X axis, and 14.5 inches in the Y axis. The throat distance is 15.5 inches and the table-to-spindle distance is 7 inches. The Interact 4 has a maximum load of 1,000 pounds, a working surface of 38 inches by 15 inches, and a positioning speed of 200 inches per minute. The accuracy rating is plus or minus 0.0008 inches and the repeatability rating is plus or minus 0.0003 inches.

Figure 4-9 is a Cincinnati Milacron 2EF CINTERNAL multi-surface grinding machine. The 2EF CINTERNAL occupies 113.625 inches by 166.625 inches of floor area, is 65 inches high, and weighs 12,500 pounds. It has a maximum swing capacity of 20 inches in the guard and a maximum swing of 13 inches in the swivel. The travel capabilities are 16 inches for the workslide, 12 inches in the longitudinal bridge adjustment, and 20 inches for table travel. The maximum grinding stroke is 17 inches and the maximum load capacity is 1,000 pounds. The axes speed for grinding is 99.9 inches per minute. The 2EF CINTERNAL has an accuracy rating of 0.000025 inches in axes resolution and 0.000030 inches in work head rotation.

There are many other CNC machines from other vendors. However, those profiled herein are representative of the types of machines used in modern manufacturing settings.

CNC INPUT MEDIA

All of the CNC machines shown in Figures 4-5 through 4-9 have controllers (Chapter 2). There are as many different controllers as there are CNC machines. Regardless of the type of controller, there must be some means of storing and inputting programs which instruct the machines. These means are called input media.

The most frequently used types of input media are punched tape and magnetic tape. Punched tape may be either the paper or mylar (plastic) variety. Mylar is more widely used because it is less prone to damage.

With earlier CNC machines a tape punch was used to put holes in a tape that was fed through it. The holes represented the program code. The punched tape was run through a tape reader that sensed the holes and fed the code into the controller. A problem with this type of input media is that it is easy to make an error when punching the tape and difficult to correct it.

An improvement to the tape punch-tape reader controller method of input is to interface a tape punch directly with a microcomputer. The code is typed via the microcomputer's keyboard. As the code is entered, it is stored so that it can be checked for accuracy and edited. Once correct, it is fed to the tape punch and the tape is prepared. This solves the error correction problem.

Magnetic tape is now more popular than punched tape. With this type of medium, the program code is affixed to the tape as magnetic spots rather than holes punched through it. Because this type of medium is becoming so popular, standards for format and coding have been developed by the Electronics Industries Association (EIA).

CNC TAPE FORMATS AND CODE STANDARDS

The most commonly used tape format is the standard RS-274 format developed by the EIA. The RS-274 is also referred to as the "word address format." In this format, program code is affixed to the tape in program lines called "blocks."

The program code conforms to one of two code standards: RS-274 or RS-358. Figure 4-10 illustrates the RS-274 tape code. This standard is the earlier of the two. It allows alphanumeric characters to be translated into binary code.

Figure 4-11 illustrates the RS-358 tape code. This is a newer approach that is less limited than the RS-274. While the RS-274 was developed primarily for numerical control applications, the RS-358 can be used in a variety of applications.

The RS-358 is based on the American Standard Code for Information Exchange (ASCII). Both codes are still used in CNC settings.

ADVANTAGES AND DISADVANTAGES OF CNC

CNC is the modern, most technologically advanced method of controlling manufacturing machines. However, it is not the best control method in every case. Like any technological development, CNC has its advantages and disadvantages. There are times when the traditional manual approach is better. CNC is indicated when one or more of the following factors is important:

- Increased productivity and accuracy
- Decreased labor and production costs

When CNC is indicated, there are advantages and disadvantages with which students of CAD/CAM should be familiar. The advantages include

1. better production and quality control;
2. increased productivity, flexibility, accuracy, and uniformity;
3. reduced labor, production, tool, and fixture costs; and
4. less parts handling and tool storage.

There are other advantages of CNC. However, these are the most important. There are also disadvantages associated with CNC. Those most frequently stated are

1. high "up-front" costs,
2. higher operating costs,
3. retraining costs, and
4. potential personnel problems.

The advantages of CNC outweigh the disadvantages. But the list of disadvantages does make the point that CNC is not always the appropriate choice.

TRACK NUMBER

8	7	6	5	4		3	2	1	
		●			●				0
					●			●	1
					●		●		2
			●		●		●	●	3
					●	●			4
			●	●	●			●	5
			●	●	●		●		6
					●	●	●	●	7
				●	●				8
			●		●			●	9
	●	●			●			●	a
	●	●			●		●		b
	●	●	●		●		●	●	c
	●	●			●	●			d
	●	●	●		●	●		●	e
	●	●	●		●	●	●		f
	●	●			●	●	●	●	g
	●	●		●	●				h
	●	●	●	●	●			●	i
	●		●		●			●	j
	●		●		●		●		k
	●				●		●	●	l
	●		●		●	●			m
	●				●	●		●	n
	●				●	●	●		o
	●		●		●	●	●	●	p
	●		●	●	●				q
	●			●	●			●	r
		●	●		●		●		s
		●			●		●	●	t
		●	●		●	●			u
		●			●	●		●	v
		●			●	●	●		w
		●	●		●	●	●	●	x
		●	●	●	●				y
		●		●	●			●	z
	●	●		●	●		●	●	(Period)
		●	●	●	●		●	●	(Comma)
		●	●		●			●	/
	●	●	●		●				+
	●				●				-
			●		●				Space
	●	●	●	●	●	●	●	●	Delete
●									CARR. RET. (EOB)
		●		●	●		●		Back Space
		●	●	●	●	●	●		Tab
			●		●		●	●	End of Record
					●				Tape Feed Hole
									Blank Tape

● = Hole in Tape

● = Tape Feed Hole

FIGURE 4-10 **RS-274 tape code.** (From COMPUTER NUMERICAL CONTROL: CONCEPTS AND PROGRAMMING, 1986, Seames, Delmar Publishers Inc.)

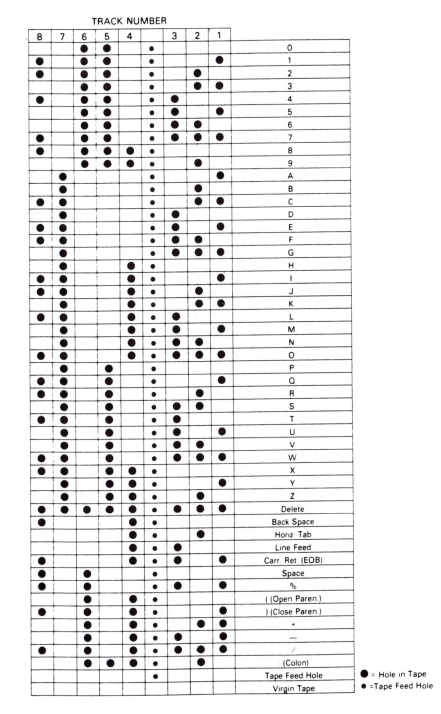

FIGURE 4-11 RS-358 tape code. (From COMPUTER NUMERICAL CONTROL: CONCEPTS AND PROGRAMMING, 1986, Seames, Delmar Publishers Inc.)

CNC APPLICATIONS

CNC applications can be viewed at two levels. First, there is the industry level. At this level, all of the various manufacturing industries use CNC machines. These include

- aerospace manufacturers
- electronics manufacturers
- automobile, truck, and bus manufacturers
- appliance manufacturers
- tooling manufacturers
- locomotive/train manufacturers

The second application level to consider is the machine level. At this level, CNC applications include

- cleaning and finishing applications
- material removal applications
- presswork and forming applications
- inspection/quality control applications

Cleaning and finishing machines perform such tasks as washing, degreasing, blasting, lapping, grinding, and deburring. There are currently several CNC cleaning and finishing machines on the market.

Material-removal machines include mills, lathes, grinders, drills, and saws. All such machines may be CNC controlled. In fact, this is the machine level application area most strongly associated with CNC.

Presswork and forming machines perform such tasks as pressing, stamping, blanking, punching, bending, and swagging. There are currently several CNC presswork and forming machines on the market.

Inspection and quality control machines perform such tasks as measuring, gaging, testing, and weighing. There are several CNC inspection and quality control machines on the market.

CNC SYSTEMS

A CNC machine is not a single unit. Rather, it is made up of several interdependent systems. Students of CNC should be familiar with the various types of CNC systems. These include control, feedback, movement, and positioning systems.

Types of CNC Control Systems

There are a number of parallels between robotics and CNC. One of the parallels is the types of control systems. As with robots, CNC machines use one of two types of control systems. These are known as either *Point-to-point* or *Continuous-path.*

Point-to-point control is the simpler, less sophisticated, and, in turn, less expensive of the two. With this type of control system, machine motion is defined by a series of straight lines (Figure 4-12). Notice in Figure 4-12 that with point-to-point control even curves must be accomplished by a series of straight lines. This limits the flexibility of point-to-point machines.

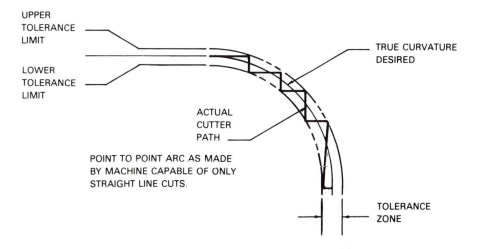

FIGURE 4-12 Point-to-point control of arcs. (From COMPUTER NUMERICAL CONTROL: CONCEPTS AND PROGRAMMING, 1986, Seames, Delmar Publishers Inc.)

Continuous-path control is more complex, sophisticated, and flexible than point-to-point control. Because of this, continuous-path CNC machines tend to be more expensive than point-to-point machines. However, advances in electronic technology are rapidly changing this. As a result, the more modern CNC machines tend to use continuous-path control. A continuous-path machine is able to accomplish angles and smooth curves (Figure 4-13).

CNC Feedback Systems

Another parallel between robots and CNC machines is in the types of electronic feedback systems used. As with robots, CNC machines use one of two types of electronic feedback systems: open-loop and closed-loop systems. Feedback systems send electronic signals to and from the motors which drive CNC machines.

Open-loop systems are the less sophisticated of the two. With this type of system, the input media carries instructions into the reader.

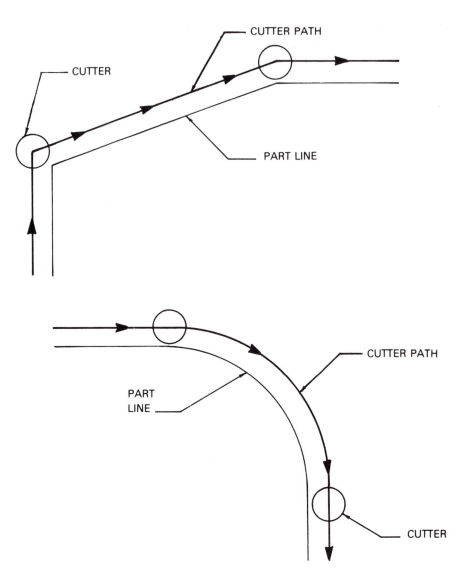

FIGURE 4-13 Continuous-path control of arcs. (From COMPUTER NUMERICAL CONTROL: CONCEPTS AND PROGRAMMING, 1986, Seames, Delmar Publishers Inc.)

The reader interprets the instructions and sends them to the storage unit. From the storage unit the instructions, in the form of electronic signals, are sent to the motor controller. From the motor controller the signals go to the drive motor which actually drives the machine in performing the task specified in the instructions (Figure 4-14).

The only feedback of which such a system is capable is from the drive motor to the memory. This is a signal which indicates that the motor has carried out the instruction received and is ready to receive another.

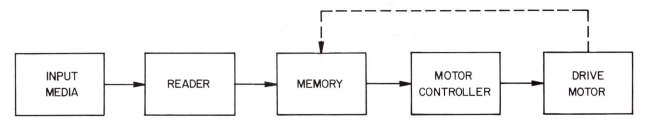

OPEN-LOOP SYSTEM

FIGURE 4-14 Open-loop feedback system.

There is no monitoring of the machine to ensure that the instructions are being properly carried out. This is the major weakness of an open-loop feedback system.

A closed-loop feedback system is more sophisticated. It is similar to an open-loop system in that instructions come from the input media through the reader and into memory. At this point, however, some important differences exist.

A closed-loop system has two additional components not available in an open-loop system: a comparison device and a feedback device (Figure 4-15). The electronic signals which come from the memory unit to the motor controller first pass through the comparison unit. Then, they go to the motor controller and on to the drive motor.

As the motor drives the machine through the required task, the comparison unit compares what the motor does with what it is supposed to do. The feedback unit allows for continuous monitoring of this nature.

With a closed-loop system, minor positioning errors caused by the drive motor can be corrected as the machine continues to run. Large errors cause the machine to shut down so that corrections can be made.

From this, the advantage of the closed-loop over the open-loop system can be seen. With an open-loop system, the machine will continue as instructed even if the drive motor is causing errors. Any part produced under these circumstances will probably be rejected.

The disadvantage of the closed-loop system is that the additional components make it more expensive than the open-loop system.

Open-loop systems use stepper drive motors. Closed-loop systems use AC, DC, or hydraulic servodrive motors. Servodrives are more sophisticated than stepper motors. However, technological developments in the area of stepper motors are making them almost as accurate as servodrives. As stepper motor technology continues to improve, open-loop systems will, in turn, become more accurate.

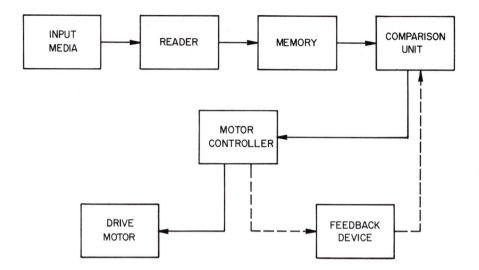

FIGURE 4-15 Closed-loop feedback system.

CNC Movement Systems

Another parallel between robots and CNC machines is in the area of movement systems. Like robots, CNC machine movement occurs within the Cartesian Coordinate System. A two-axis movement system allows movement along X and Y axes; a three-axis movement system allows movement along X, Y, and Z (Figure 4-16).

Movement within a given system is accomplished by assigning X and Y, or X, Y, and Z coordinates. Notice on Figure 4-17 that two intersecting axes divide the movement area into four quadrants (quadrants I, II, III, and IV, labelled counterclockwise, beginning in the upper right-hand quadrant). The point of origin, or 0,0 point, is the point where the X and Y axes intersect.

Any X coordinate that falls to the right of the origin (or, in other words, in quadrant I or IV) is a positive X coordinate. Any X coordinate that falls to the left of the origin (in quadrant II or III) is a negative X coordinate. Any Y coordinate above the origin (quadrants I or II) is a positive Y coordinate. Any Y coordinate below the origin (quadrants I and II) is a negative Y coordinate.

Figure 4-17 is a two-axis system. Were it a three-axis system, it would have an additional axis (the Z axis). Imagine this axis as running in and out perpendicular to the page. Any Z coordinate behind the origin is a negative coordinate. Any Z coordinate in front of the origin is a positive coordinate.

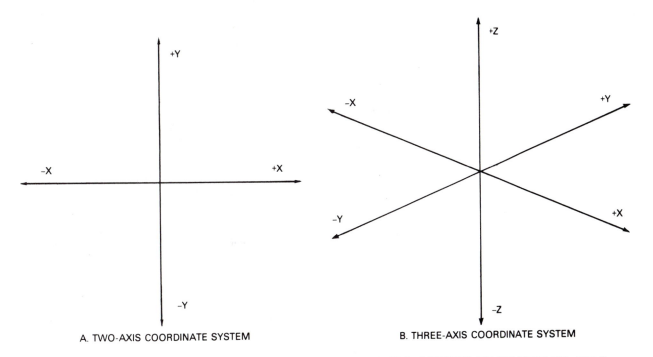

A. TWO-AXIS COORDINATE SYSTEM   B. THREE-AXIS COORDINATE SYSTEM

FIGURE 4-16 Three-axis movement system. (From COMPUTER NUMERICAL CONTROL: CONCEPTS AND PROGRAMMING, 1986, Seames, Delmar Publishers Inc.)

CNC Positioning Systems

CNC machines must position their tools within the movement systems described in the previous section. There are two types of positioning systems for accomplishing this: *Absolute* positioning and *Incremental* positioning.

With *absolute* positioning, all machine positions start at the origin or 0,0 point (Figure 4-18).

With *incremental* positioning, the origin or 0,0 point moves as the machine moves. The first point is measured from the origin, the second point from the first, the third from the second, and so on.

CNC PROGRAMMING COORDINATES

CNC programming personnel must be able to assign coordinates as the first step in developing programs. The coordinates are used as the basis of the programs which instruct CNC machines in accomplishing their tasks. Programming coordinates are assigned in either the absolute or incremental positioning systems.

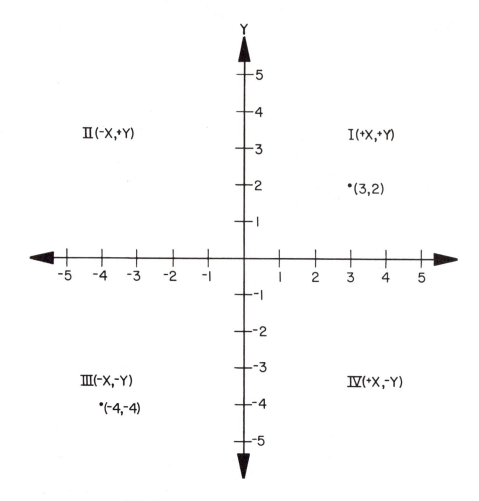

FIGURE 4-17 Quadrants in a coordinate system.

To understand how programming coordinates are assigned in the absolute positioning system, examine Figure 4-18. In this figure, there is a plate through which two holes are to be drilled. The origin or 0,0 point is the lower left-hand corner of the plate.

Using the dimensions shown, programming coordinates can be determined for Holes 1 and 2. The coordinates for Hole 1 are

X1.50, Y1.50

The coordinates for Hole 2 also begin at the 0,0 point. They are

X3.75, Y1.50

Notice from this example that the coordinates for Hole 2 are independent of those for Hole 1. This is the fundamental characteristic of programming coordinates in the absolute positioning system.

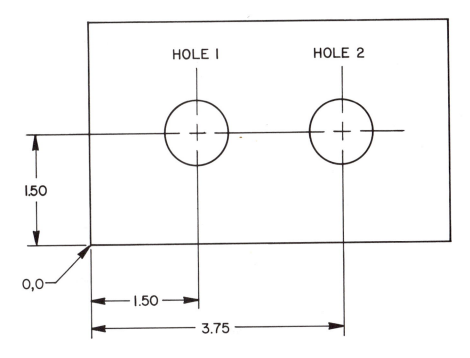

FIGURE 4-18 Absolute positioning.

To understand how programming coordinates are assigned in the incremental positioning system, return to Figure 4-18. In this system the position of Hole 2 is relative to that of Hole 1. Each time the machine moves to an assigned position, that position becomes the 0,0 point. The next coordinates are relative to this new point of origin.

Assuming the lower left-hand corner to be the machine's position at the beginning of the program, the coordinates for Holes 1 and 2 can be determined. The coordinates for Hole 1 are

X1.50, Y1.50

As the machine moves to this point, it becomes the new 0,0 point. The coordinates for Hole 2 are

X2.25, Y1.50

The X coordinate is the distance between Hole 1 and Hole 2 (3.75 − 1.50 = 2.25).

WRITING CNC PROGRAMS

Another parallel between robots and CNC machines is in the area of programming. In robotics, there are almost as many programming languages as there are robots. CNC machines are more standardized than this. However, there is a lot of variety in how they are programmed. In fact, no two machines are programmed exactly alike.

In spite of this, there are points of commonality that can be generalized to all CNC machines. Students of CNC should be familiar with these. There are a number of programming formats used in CNC. Two of the more commonly used formats are the *Machinist Shop Language* format and the *Word Address* format.

Machinist Shop Language

Machinist Shop Language is a conversational format that uses widely understood shop words. The CNC machine's controller converts the words in an MSL program into the proper machine codes.

Each line in an MSL program is called an "event." There are two types of events: *Motion* events and *Non-motion* events.

Motion events move the machine to the desired locations specified by coordinates. Non-motion events allow the machine to perform the required task at each location (i.e., drill, cut, etc.). Each event in an MSL program is written using certain commands, preparatory functions or G Codes, variable codes or V Codes, and auxiliary or AUX codes.

In his book *COMPUTER NUMERICAL CONTROL: Concepts and Programming* (Delmar Publishers, 1986), Warren S. Seames summarizes the more commonly used MSL commands and codes as follows:

Machinist Shop Language Commands

This is a list of commonly used Machinist Shop Language commands:

A (absolute)—Specifies absolute positioning.

ARC—If used by itself, institutes the cutting of an arc. If used with CW or CCW, tells the computer that an arc is to be cut in a clockwise or counterclockwise direction. Following an ARC/direction command, the computer will look for information describing the arc in the following two events.

AUX (auxiliary)—Allows changes to be made in normal control functions. For example, the direction of the X, Y, and Z axes may be changed, and mirror imaging may be instituted. AUX codes act like the miscellaneous functions in a word address format.

CALL—Executes a subroutine. For example, CALL 1 tells the machine to carry out the instructions in word address format.

CCW—Specifies a counterclockwise arc rotation.

CW—Specifies a clockwise rotation.

DO—DO loop. Anything that is repeated over equal intervals of space (a row of holes, for example) may be placed in a DO loop. DO 5 tells the machine to perform a specified operation five times.

DWELL—Halts execution of the program until the START button is manually depressed. When START is pressed, the program continues, starting at the next event.

END—There are three uses for the END command. (1) In a DO loop, END signals the end of the loop. (2) In a subroutine, END signals the end of the subroutine. (3) In a program, END signals the end of the program.

F (feed)—Tells the machine to make tool movements at the programmed feedrate.

FEED—Assigns a feedrate.

G (G code)—A preparatory function, G code calls up certain "canned" or standard cycles contained within the computer for such operations as drilling, boring, and reaming.

I (incremental)—Specifies incremental positioning.

R (rapid)—Tells the machine to make tool movements at rapid traverse.

SUBR (subroutine)—Like a miniprogram within a program, sections of a program that are to be repeated are often placed in a subroutine to eliminate having to program the same information twice. The subroutine is instituted by using the CALL command.

TOOL—Like a dwell, halts the program so that a tool can be inserted in the spindle. If the machine is equipped with three axes, TOOL also acts to assign certain tool length and/or cutter diameter values.

V (variable)—Assigns values to certain program variables such as canned-cycle feedrates and feed-engagement points.

MSL Preparatory Functions (G Codes)

The following is a list of preparatory functions used in conjunction with Machinist Shop Language:

G40—Cutter diameter compensation cancel.
G41—Cutter diameter compensation left.
G42—Cutter diameter compensation right.
G51—Institute polar rotation.
G52—Polar rotation cancel.
G53—Institute scaling.
G54—Scaling cancel.
G76—Hole milling.
G77—Circular pocket milling.
G78—Rectangular pocket milling.
G79—Bolt circle pattern.
G80—Canned cycle cancel.

G81—Basic drilling cycle.

G82—Counter-boring/spot-facing (feed in, timed dwell, rapid out).

G83—Peck drilling cycle (feed in, rapid out, feed in, etc.).

G85—Boring cycle (feed in, feed out).

G86—Boring in one direction cycle (feed in, rapid out).

G87—Chip-breaking cycle (feed in, retract .050, feed in, etc.).

G89—Flat bottom boring cycle (feed in, timed dwell, feed out).

MSL Variable (V) Codes

V11—X-axis polar center (must be absolute dimension).

V12—Y-axis polar center (must be absolute dimension).

V13—Polar rotation index angle (must be incremental). A negative number indicates clockwise rotation, a positive number indicates counterclockwise rotation.

V14—Radius for polar moves (value must be positive).

V15—Angle for polar moves or angle of first hole in a bolt circle pattern.

V16—Angle of last hole in a bolt circle pattern or X-axis scaling value.

V17—Number of holes to be machined in a bolt circle or Y-axis scaling value.

V18—Diameter of bolt circle or Z-axis scaling value.

V20—Feedrate for G80 series canned cycles. Must be .100 for G83 or G87.

V21—Buffer zone for G80 series canned cycles. Must be .100 for G83 or G87.

V22—Dwell time when using G82 or G89.

V23—Maximum peck when using G83 or G87.

V40—Z-axis start height for pecked milling.

V41—Length of pocket on X axis (must be incremental).

V42—Width of pocket on Y axis (must be incremental) or number of rotations for helical interpolation.

V43—Depth of pocket on Z axis.

V44—Pocket corner radius or diameter of circle if circular pocket milling.

V45—Stepover value for pocket milling.

V46—Maximum depth of cut.

V47—Stock left for finish pass.

V48—Finish pass feedrate.

V49—Tool diameter for pocket milling (cutter compensation cannot be active for pocket milling).

MSL Auxiliary (AUX) Codes

The following is a complete list of auxiliary codes commonly used in Machinist Shop Language.

AUX 100—Reverses sign of X axis.

AUX 200—Reverses sign of Y axis.

AUX 300—Reverses sign of X and Y axes.

AUX 400—Reverses sign of Z axis.

AUX 500—Reverses sign of X and Z axes.

AUX 600—Reverses sign of Y and Z axes.

AUX 700—Reverses sign of X, Y, and Z axes.

AUX 800—Turns off mirror image.

AUX 1000—Causes machine to continue to the next move before reaching its target (used only with contouring operations).

AUX 1101—Absolute zero shift.

AUX 1110—Turns off software limits.

AUX 1111—Turns on software limits.

AUX 1400—Feed percentage override for feedrate moves.

AUX 1401—Feed percentage override for feed and rapid moves.

AUX 1900—Single-step event mode.

AUX 1901—Single-step axis movement mode.

AUX 2000—Cancels AUX 1000.

AUX 2500—Sets control to use Z axis.

AUX 2600—Sets control to allow manual use of Z axis.

Figure 4-19 is an example of an MSL program written for a two-axis drilling operation. Compare the language in the program with the command and code explanations in the previous sections. A close examination will reveal that three ⅜-inch diameter holes are to be drilled through a workpiece. The holes have been assigned absolute programming coordinates.

```
XO/YO = LOWER LEFT CORNER OF PART
TOOL CHANGE = X-2 Y-1.5
SPINDLE SPEED = 2500 RPM

1   X-2 Y-1.5   R A
2   TOOL 1                REM: 3/8 DRILL
3   X.5 Y2      R A
4   DWELL                 REM: DRILL HOLE
5   X1.25 Y1.5 R A
6   DWELL                 REM: DRILL HOLE
7   X2 Y.5      R A
8   DWELL                 REM: DRILL HOLE
9   X-2 Y-1.5   R A
10  END
```

FIGURE 4-19 **Sample MSL program.** (From COMPUTER NUMERICAL CONTROL: CONCEPTS AND PROGRAMMING, 1986, Seames, Delmar Publishers Inc.)

```
XO/YO = LOWER LEFT CORNER OF PART
TOOL CHANGE = X-2 Y-1.5
TOOL 3/8 DRILL SPINDLE SPEED 2500 RPM

N010 G00 G70 G90 X-2 Y-1.5  M06    REM:3/8
DRILL
N020 X.5 Y2
N030 G04
N040 X1.25 Y1.5
N050 G04
N060 X2 Y.5
N070 G04
N080 X-2 Y-1.5
N090  M30
```

FIGURE 4-20 **Sample word address program.** (From COMPUTER NUMERICAL CONTROL: CONCEPTS AND PROGRAMMING, 1986, Seames, Delmar Publishers Inc.)

Word Address Format

The Word Address format is also known as the variable block format. Although this programming format was originally developed for use in situations in which tape is the input medium, it can also be used for manual input on CNC machines. With this format, each line of programming code or block can vary in length.

In earlier tape formats, zeros had to be used as place holders so that all lines or blocks were the same length. The word address format is more efficient because only the code required to specify the instructions need be included. Every program line or block can be a different length.

In his book, COMPUTER NUMERICAL CONTROL: Concepts and Programming, Warren S. Seames explains the word address format as follows:

Addresses

The block format for word address is

$$N \ldots G \ldots X \ldots Y \ldots Z \ldots I \ldots J \ldots K \ldots F \ldots S \ldots T \ldots M \ldots$$

Only the information needed on a line need be given. Each of the letters is called an address (or word). The various words are as follows:

N—Designates the start of a block. Program lines or blocks are sometimes called sequence lines. On some machinery, the address "O" may also be used to start a block of information.

G—Initiates a preparatory function. Preparatory functions change the control mode of the machine. Examples of preparatory functions are rapid/feed rate mode, drilling mode, tapping mode, boring mode, and circular interpolation. Preparatory functions are called prep functions, or more commonly, G codes.

X—Designates an X-axis coordinate. X is also used to enter a time interval for a timed dwell.

Y—Designates a Y-axis coordinate.

Z—Designates a Z-axis coordinate.

I—Identifies the X-axis location of an arc centerpoint.

J—Identifies the Y-axis location of an arc centerpoint.

K—Identifies the Z-axis location of an arc centerpoint.

S—Sets the spindle RPM.

T—Specifies the tool to be used in a tool change.

M—Initiates miscellaneous functions (M functions). M functions control auxiliary functions such as the turning on and off of the spindle and coolant.

Seames also lists and explains many of the more commonly used word address format codes used in programming CNC machines:

Preparatory Functions (G Codes) Used in Milling

The following is a list of preparatory functions commonly used in CNC:

G00—Rapid traverse positioning.
G01—Linear interpolation (feedrate movement).
G02—Circular interpolation clockwise.
G04—Dwell.
G10—Tool length offset value.
G17—Specifies X/Y plane.
G18—Specifies X/Z plane.
G19—Specifies Y/Z plane.
G20—Inch data input (on some systems).
G21—Metric data input (on some systems).
G22—Safety zone programming.
G23—Cross through safety zone.
G27—Reference point return check.
G28—Return to reference point.
G29—Return from reference point.
G30—Return to second reference point.
G40—Cutter diamond compensation cancel.
G41—Cutter diamond compensation left.
G42—Cutter diameter compensation right.
G43—Tool length compensation positive direction.
G44—Tool length compensation negative direction.
G45—Tool offset increase.
G46—Tool offset decrease.
G47—Tool offset double increase.
G48—Tool offset double decrease.
G49—Tool length compensation cancel.
G50—Scaling off.
G51—Scaling on.
G73—Peck drilling cycle.
G74—Counter tapping cycle.
G76—Fine boring cycle.
G80—Canned-cycle cancel.
G81—Drilling cycle.
G82—Counter boring cycle.
G83—Peck drilling cycle.
G84—Tapping cycle.
G85—Boring cycle (feed return to reference level).
G86—Boring cycle (rapid return to reference level).
G87—Back boring cycle.

G88—Boring cycle (manual return).
G89—Boring cycle (dwell before feed return).
G90—Specifies absolute positioning.
G91—Specifies incremental positioning.
G92—Program absolute zero point.
G98—Return to initial level.
G99—Return to reference (R) level.

Miscellaneous Functions (M Codes) Used in Milling and Turning

The following is a list of miscellaneous functions used in milling and turning programs:

M00—Program stop.
M01—Optional stop.
M02—End of program (rewind tape).
M03—Spindle start clockwise.
M04—Spindle start counterclockwise.
M05—Spindle stop.
M06—Tool change.
M08—Coolant on.
M09—Coolant off.
M13—Spindle on clockwise, coolant on (on some systems).
M14—Spindle on counterclockwise, coolant on.
M17—Spindle and coolant off (on some systems).
M19—Spindle orient and stop.
M21—Mirror image X axis.
M22—Mirror image Y axis.
M23—Mirror image off.
M30—End of program, memory reset.
M41—Low range.
M42—High range.
M48—Override cancel off.
M49—Override cancel on.
M98—Jump to subroutine.
M99—Return from subroutine.

Preparatory Functions (G Codes) Used in Turning

The following is a list of preparatory functions used in CNC milling programs:

G00—Rapid traverse positioning.
G01—Linear interpolation (feedrate movement).
G02—Circular interpolation clockwise.
G03—Circular interpolation counterclockwise.

G04—Dwell.
G10—Tool length offset value setting.
G17—Specifies X/Y plane.
G18—Specifies X/Z plane.
G19—Specifies Y/Z plane.
G20—Inch data input (on some systems).
G21—Metric data input (on some systems).
G22—Stored stroke limit on.
G23—Stored stroke limit off.
G27—Reference point return check.
G28—Return to reference point.
G29—Return from reference point.
G30—Return to second reference point.
G40—Tool nose radius compensation cancel.
G41—Tool nose radius compensation left.
G42—Tool nose radius compensation right.
G50—Programming of work coordinate system.
G68—Mirror image for double turrets on.
G69—Mirror image for double turrets off.
G70—Inch programming (some systems) or finish cycle.
G71—Metric programming (some systems) or stock removal in turning
  code.
G72—Stock removal in facing code.
G73—Pattern repeat.
G74—Z-axis peck drilling.
G75—Groove cutting cycle, X axis.
G76—Multipass thread cutting.
G90—Absolute positioning.
G91—Incremental positioning.
G94—Per minute feed (some systems).
G95—Per revolution feed (some systems).
G98—Per minute feed (some systems).
G99—Per revolution feed (some systems).

   Figure 4-19 has already been cited as an example of an MSL program written for a two-axis drilling operation. A close examination of this program revealed that it contained the instructions for drilling three ⅜-inch diameter holes. Figure 4-20 is a word address version of this program. A close examination of this program will show that it, too, contains the instructions for drilling three ⅜-inch diameter holes.

CNC-Related Math

CNC technicians frequently use both algebra and trigonometry in performing such tasks as coordinate and missing dimension calculations. Figure 4-21 summarizes some of the more frequently used terms in trigonometry. Figures 4-22 and 4-23 summarize the algebraic and trigonometric formulas most frequently used on the job in a CNC setting.

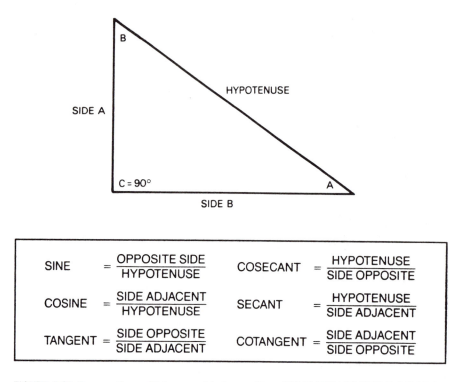

SINE	$= \dfrac{\text{OPPOSITE SIDE}}{\text{HYPOTENUSE}}$	COSECANT	$= \dfrac{\text{HYPOTENUSE}}{\text{SIDE OPPOSITE}}$
COSINE	$= \dfrac{\text{SIDE ADJACENT}}{\text{HYPOTENUSE}}$	SECANT	$= \dfrac{\text{HYPOTENUSE}}{\text{SIDE ADJACENT}}$
TANGENT	$= \dfrac{\text{SIDE OPPOSITE}}{\text{SIDE ADJACENT}}$	COTANGENT	$= \dfrac{\text{SIDE ADJACENT}}{\text{SIDE OPPOSITE}}$

FIGURE 4-21 Frequently used trigonometric terms. (From COMPUTER NUMERICAL CONTROL: CONCEPTS AND PROGRAMMING, 1986, Seames, Delmar Publishers Inc.)

SUMMARY

NC is a broad term which encompasses the traditional approach to numerical control as well as the more modern outgrowths such as computer numerical control (CNC) and direct numerical control (DNC). There are CNC machines for performing most manufacturing processes, including cleaning and finishing, inspection and quality control, pressing and forming, and material removal.

Programs for CNC machines are stored and input via some type of input medium. The most widely used media include punched

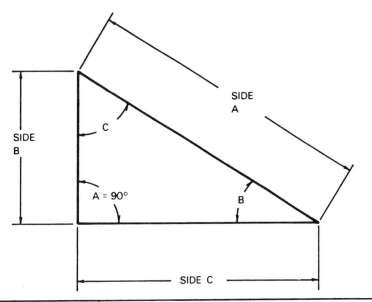

KNOWN VARIABLES	SOLUTION FORMULAS		
SIDE a, ANGLE B	b = a × SIN B	c = a × COS B	C = 90° − B
SIDE a, ANGLE C	b = a × COS C	c = a × SIN C	B = 90° − C
SIDE b, ANGLE B	$a = \dfrac{b}{SIN\ B}$	c = b × COT B	C = 90° − B
SIDE b, ANGLE C	$a = \dfrac{b}{COS\ C}$	c = b × TAN C	B = 90° − C
SIDE c, ANGLE B	$a = \dfrac{c}{COS\ B}$	b = c × TAN B	C = 90° − B
SIDE c, ANGLE C	$a = \dfrac{c}{SIN\ C}$	b = c × COT C	B = 90° − C
SIDES a AND b	$c = \sqrt{a^2 - b^2}$	$SIN\ B = \dfrac{b}{a}$	C = 90° − B
SIDES a AND c	$b = \sqrt{a^2 - c^2}$	$SIN\ C = \dfrac{c}{a}$	B = 90° − C
SIDES b AND c	$a = \sqrt{b^2 + c^2}$	$TAN\ B = \dfrac{b}{c}$	C = 90° − B

FIGURE 4-22 Frequently used formulas. (From COMPUTER NUMERICAL CONTROL: CONCEPTS AND PROGRAMMING, 1986, Seames, Delmar Publishers Inc.)

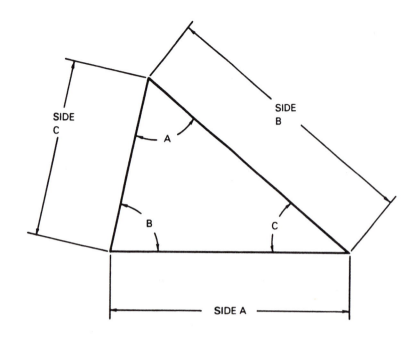

ONE SIDE AND TWO ANGLES KNOWN: GIVEN: SIDE a, OPPOSITE ANGLE A, AND OTHER ANGLE B
$C = 180° - (A + B)$ $\qquad b = \dfrac{a \times SIN\ B}{SIN\ A}$ $\qquad c = \dfrac{a \times SIN\ C}{SIN\ A}$
TWO SIDES AND THE ANGLE BETWEEN THEM KNOWN: GIVEN: SIDES a, b, AND ANGLE C
$TAN\ A = \dfrac{a \times SIN\ C}{b - (a \times COS\ C)}$ $\qquad B = 180° - (A + C)$ $\qquad c = \dfrac{a \times SIN\ C}{SIN\ A}$ $c = \sqrt{a^2 + b^2 - (2ab \times COS\ C)}$
TWO SIDES AND ANGLE OPPOSITE ONE SIDE KNOWN: GIVEN: SIDE a, OPPOSITE ANGLE A, AND SIDE B
$SIN\ B = \dfrac{b \times SIN\ A}{a}$ $\qquad C = 180° - (A + B)$ $\qquad c = \dfrac{a \times SIN\ C}{SIN\ A}$
ALL THREE SIDES KNOWN:
$COS\ A = \dfrac{b^2 + c^2 - a^2}{2\,b\,c}$ $\qquad SIN\ B = \dfrac{b \times SIN\ A}{a}$ $\qquad C = 180° - (A + B)$

FIGURE 4-23 Frequently used formulas. (From COMPUTER NUMERICAL CONTROL: CONCEPTS AND PROGRAMMING, 1986, Seames, Delmar Publishers Inc.)

tape, which may be paper or plastic, and magnetic tape, which is the more modern medium. Instructional codes on CNC media conform to one of two standards: RS-274 or RS-358.

CNC offers advantages and has some disadvantages. The advantages are better production and quality control; increased productivity, flexibility, accuracy, and uniformity; reduced labor, production, tool, and fixture costs; and less parts handling and tool storage. The disadvantages of CNC are high up-front costs, higher operating costs, retraining costs, and potential personnel problems.

CNC applications can be viewed at two levels: the industry level and the machine level. At the industry level, CNC applications include aerospace, electronics, automobile, truck, bus, locomotive, appliance, tooling, and mechanical manufacturing. At the machine level, CNC applications include cleaning, finishing, material removal, presswork, forming, inspection, and quality control machines.

CNC machines are actually systems made up of smaller, interdependent systems, including control, feedback, movement, and positioning systems. In addition to these systems, there are two systems used for assigning programming coordinates in CNC: absolute and incremental. In the absolute system, all coordinates are measured from the original 0,0 point. In the incremental system, each successive coordinate is measured from the last coordinate. Once program coordinates have been assigned, CNC programs can be written. They are written in a variety of formats. Two of the more common formats are the Machinist Shop Language and Word Address formats.

Chapter Four REVIEW

1. Define the term "numerical control."
2. Who is the father of NC?
3. List the four categories of manufacturing machines which might be numerically controlled.
4. What are the drawbacks to punched tape as an input medium?
5. Explain the two most widely used tape format standards.
6. List two advantages of CNC.
7. List two disadvantages of CNC.
8. What types of industries use CNC machines?
9. List and explain four types of CNC systems.

Several different manufacturing processes, including materials handling and assembly operations, can be automated through the use of industrial robots. When manufacturing is fully integrated, industrial robots will represent one or more components in an overall CIM system.

Major Topics Covered

- Overview of Industrial Robots
- Categories of Robots
- Robot End Effectors
- Robot Sensors
- Robot Programming
- Robots and People
- Robotics-Related Math

Chapter Five

Industrial Robots

When the Diecast Division of Tecumseh Products Company, a major small-engine manufacturer, began planning increased automation, it did so with specific goals in mind.

The company was already using automatic extractors on seven of the thirty-four die casting machines at its 150,000-square-foot plant, freeing operators from the unpleasant and potentially hazardous task of manually unloading 700°F castings. But in order to achieve the productivity increase desired, the new round of automation would have to accomplish far more than simple part extraction or single-machine tending.

Working with ASEA ROBOTICS, Inc., of New Berlin, Wisconsin, Tecumseh developed a multifunction automated cell that produces sixty aluminum engine castings per hour, a 50% productivity increase from the forty castings possible using manual methods.

Robot Controls the Process

At the center of the four-machine cell is an ASEA IRB 60/2 robot, mounted on a 9½-inch riser. The ASEA S2 robot controller also serves as the cell controller, accepting eight inputs from the various machines and sending three outputs to initiate cycles.

The cell operates twenty-four hours a day, five days a week. The only human interventions required are unloading bins of completed castings and random quality control checks.

The cycle begins when an 800-ton automatic die casting machine, made by Prince Corporation, signals the robot that it has a completed part in its die. At this point, the robot, having verified that the die is open, reaches in and extracts the part. A sensor on the robot's gripper confirms that the part has been securely grasped. As the robot leaves the die casting machine, it signals the machine, through a hard-wired limit switch, to begin another casting operation.

Courtesy of ASEA Robotics, Inc., Hue Schlagel.

The robot then moves the hot casting to a radiator-cooled quench tank, where the part's temperature will be reduced to about 200°F in preparation for trimming. The quench tank also contains an anticorrosion chemical.

The next step, which requires the plus-or-minus 0.016-inch repeatability of the IRB 60/2 robot, involves loading an E.A. Doyle trim press. After verifying that the trim press is open and that there is no part in the press, the robot loads the press and signals it to start.

When the trimming cycle is complete, the robot extracts the casting. Here, however, it must use the opposite side of its custom-designed double gripper. The reason is that the runner that had provided an appropriate gripping surface while the part was being extracted from the die casting machine has been trimmed off in the press. Therefore, the gripping surface must now be on the edge of the trimmed casting.

The final process involves the robot loading the trimmed casting into a multiple-spindle drilling fixture. A signal from the robot initiates the drilling, and the retracting spindles signal the robot to remove the part, which is placed on a 5½-foot conveyor that carries the casting out of the cell and deposits it in a wire bin.

The bins hold about 150 castings each and are replaced with empty bins as needed. From the cell, the castings are taken to a Goff barrel-blast machine for shot-blasting, then packed for shipment.

Timing Critical

In order for the cell to attain maximum efficiency, Tecumseh had determined that the entire process should take no longer than 65 to 70 seconds. In fact, the cell completes a cycle in exactly 60 seconds.

The elapsed time between the die casting machine's signalling "die open" and the robot's completely unloading the machine and tripping a limit switch is just 7.88 seconds. The quenching process and trim press

(continued)

loading take 20.21 seconds. The trim press cycle and unloading, plus the drill press loading, together take 19.14 seconds.

From drill press cycle start to gripper unload at the conveyor takes only 4.06 seconds. The robot then spends 7.51 seconds to return to home position and is at rest for 1.20 seconds before receiving a signal from the die casting machine to unload another casting.

A major feature that allows cell cycle times to remain low is the robot's double-gripper and effector, engineered and built by ASEA ROBOTICS. If the robot were to require a complete gripper exchange, the cycle time of 60 seconds would not be possible.

Instead, the robot's end effector includes two complete grippers opposite one another. The robot's five-axis wrist motion moves the 24-inch-long end effector up to 180 degrees in either direction, depending upon which gripper assembly is needed. A 90° rotary actuator built into the end effector provides an economical sixth axis of motion. This allows the end effector to shift the part from a vertical position as it exits the trim press to a horizontal position for drilling.

In addition, the end effector's length, plus the robot's riser mounting, permit easy access to the automatic die casting machine. The robot must reach 87¾ inches from its centerline in order to extract the castings.

Safety Stressed

During production, the cell is completely automatic; the die casting machine includes automatic ladling and spraying, and the robot controller directs the other processes. The only manual operation is removing bins as they become full of castings.

Because of its automatic nature—and the desire to take advantage of the increased safety this can allow—the entire cell, with the exception of the robot control cabinet, is enclosed by a 7½-foot chain-link fence. The entry gate is wired with a positive safety stop that brings all robot operations to a halt if the gate is opened.

This can be overridden during robot programming only. However, when the programming unit is removed from the ASEA control cabinet, the robot will not operate at more than 25% of maximum speed, thus providing an additional margin of safety.

An access port cut into the fence at the rear of the cell allows the robot's programming unit to be brought conveniently into the machine area without the dangerous practice of passing cable over the top of the fence.

Additional safety features to protect not only workers but machines are provided in the form of sensors that confirm that parts are securely gripped and properly loaded in machines.

Flexibility Important

Because the demand for engine castings is seasonal—for example, snowmobile engines are made in the summer, garden tractor and lawnmower engines in the winter—a flexible automated cell was essential. The easily programmed ASEA S2 controller, plus the versatile and reliable IRB 60/2 robot, allow this flexibility.

Although the robot-tended cell has so far produced primarily castings used on 12-horsepower garden tractor engines, there have been tests run on other castings as well.

These have demonstrated that robot programs can be easily and quickly modified to accommodate different-sized castings. The robot's weight-handling ability is not a concern: the castings weigh in the 5- to 7-pound range, while the robot has a rated handling capacity of 132 pounds, more than enough to handle the gripper and part.

Castings from Tecumseh Diecast Division are shipped to the Lauson Division of Tecumseh, in nearby New Holstein, Wisconsin, where all of Tecumseh's four-cycle engines are assembled.

(continued)

Results Impressive

Careful planning, cost-justification, and the selection of a robot company with broad application experience have been major factors in the success of flexible automation at Tecumseh Products' Diecast Division.

The robot-tended cell is, in one compact operation, accomplishing the work that had been done manually in three areas of the plant—die casting, trimming, and drilling. Cost savings have been substantial.

Payback is expected to occur more rapidly than the eighteen months originally projected. Increased productivity and improved quality will account for the entire payback. Not only is the automated cell making more castings in less time, but workers who would have had to tend the cell's machines were shifted to other, more productive tasks.

The industrial robot is one of the most important developments in the history of manufacturing technology. Most people view robotics as a relatively new field when, in reality, this technology is over twenty years old. However, like so many other forms of automation, the industrial robot did not begin to see wide-scale applications until the development of the microprocessor. The microprocessor is the enabling device which led to the development of highly capable, affordable industrial robots. Economics provided the impetus for wide-scale installation of robots.

OVERVIEW OF INDUSTRIAL ROBOTS

Before examining the historical development of robots and the rationale for this development, the reader should understand what a robot is, as well as what it isn't.

A Definition of Robots

As is sometimes the case with new and emerging technological developments, there are a variety of definitions used for the term "robot." Depending on the definition used, the number of robot installations in this country and others will vary widely. There are a variety of single-purpose machines used in manufacturing plants which, to the lay person, would appear to be robots. These machines are hard-wired to perform one single function. They cannot be reprogrammed to perform a different function. These single-purpose machines do not fit the definition for industrial robots that is coming to be widely accepted. This definition is the one developed by the Robot Institute of America (RIA). The RIA's definition for a robot is:

> A robot is a reprogrammable multifunctional manipulator designed to move material, parts, tools, or specialized devices through variable programmed motions for the performance of a variety of tasks.

Notice that the RIA's definition contains the words "reprogrammable" and "multifunctional." It is these two characteristics which separate the true industrial robot from the various single-purpose machines used in modern manufacturing firms. The term "reprogrammable" implies two things: that the robot operates according to a written program, and that this program can be rewritten to accommodate a variety of manufacturing tasks. The term "multifunctional" means that the robot is able, through reprogramming and the use of a variety of end effectors, to perform a number of different manufacturing tasks. Definitions written around these two critical characteristics are becoming the accepted definitions among manufacturing professionals. Figures 5-1 and 5-2 are examples of modern industrial robots which fit this definition.

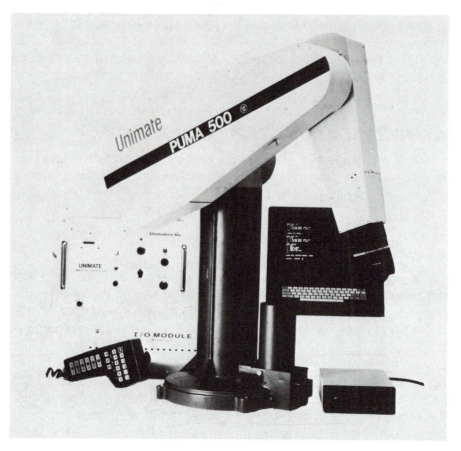

FIGURE 5-1 Puma 500 industrial robot. Courtesy of UNIMATION Incorporated, a Westinghouse Company.

Historical Development of Robots

The history of robots is as old as man's attempts to automate manufacturing. Although robots which fit the definition presented in the preceding section have been in existence for only about twenty years, students of CAD/CAM should be familiar with some of the historical developments which led to the current state of industrial robotics.

The early 1800s saw the development of one of the first industrial robots, a programmable loom used in the textile industry. Between the 1800s and the early 1950s, a variety of automated manufacturing machines and devices were developed. In 1892, a motorized crane which used a special gripper to remove white-hot steel ingots from a blast

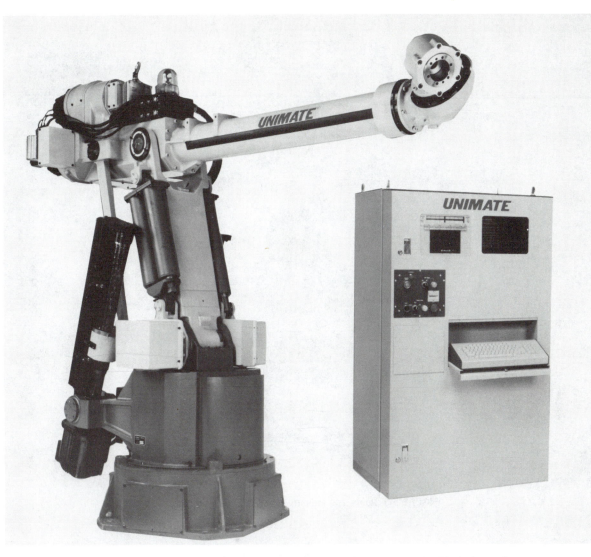

FIGURE 5-2 Unimate industrial robot. Courtesy of UNIMATION Incorporated, a Westinghouse Company.

furnace was developed by Seward Babbitt. In 1938, a programmable spray painting machine was developed for use by the DeVilbiss Company.

The first articulated arm came about in 1951 and was used by the Atomic Energy Commission. Then, in 1954, the first programmable robot was developed. The designer of this revolutionary machine was George Devol.

Devol coined the buzzword "universal automation," which was later shortened to "unimation" and became the name of the first commercial industrial robot vendor. Figures 5-3 and 5-4 are examples of modern industrial robots produced by Unimation.

FIGURE 5-3 Puma 700 industrial robot. Courtesy of UNIMATION Incorporated, a Westinghouse Company.

The first commercially produced robot was developed in 1959. Then, in 1962, the first industrial robot to be used on a production line was installed by General Motors. This robot was produced by Unimation. A major step forward in robot control occurred in 1973 with the development of the T-3 industrial robot by Cincinnati Milacron. The T-3 robot was the first commercially produced industrial robot controlled by a minicomputer. Cincinnati Milacron is still a leading vendor of industrial robots.

Since the late 1960s and early 1970s, the number of robot installations in manufacturing firms has grown rapidly. Figure 5-5 is a chart which indicates the current and projected growth of industrial robot installations in the United States through the year 1990. The data were collected and projected by the Industrial Robot Division of Cincinnati Milacron.

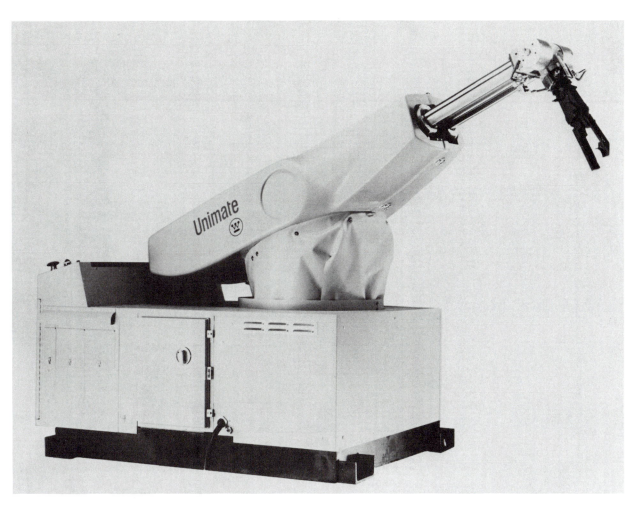

FIGURE 5-4 Unimate industrial robot. Courtesy of UNIMATION Incorporated, a Westinghouse Company.

Notice that in 1965 there were well under 5,000 industrial robot installations. By 1985, this number had grown to 15,000. But the most rapid growth is projected to take place between 1985 and 1990, when, at the turn of the decade, there will be over 100,000 industrial robots installed in manufacturing plants in this country.

Rationale for Industrial Robots

You learned earlier that the microprocessor was the enabling device with regard to the wide-scale development and use of industrial robots. But major technological developments do not take place simply because of a new capability. Something must provide the impetus for

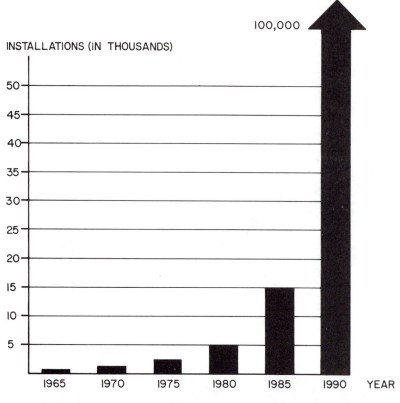

INSTALLATIONS (IN THOUSANDS)

100,000

FIGURE 5-5 Growth of industrial robot installations in the United States.

taking advantage of the new capability. In the case of industrial robots, the impetus was economics.

First, the rapid inflation of wages experienced in the 1970s increased the personnel costs of manufacturing firms tremendously. At the same time, foreign competition became a serious problem for American manufacturers. Foreign manufacturers who had undertaken automation on a wide-scale basis, such as Japan, began to gain an increasingly large share of the American market for manufactured goods, particularly automobiles.

Through a variety of automation techniques, including robots, Japanese manufacturers, beginning in the 1970s, were able to produce better automobiles more cheaply than non-automated American manufacturers who were continually engaged in struggles with organized labor. Consequently, in order to survive, American manufacturers were forced to consider any technological developments that could help improve productivity.

It became imperative to produce better products at lower costs in order to be competitive with foreign manufacturers. Other factors such as the need to find better ways of performing dangerous manufacturing tasks contributed to the development of industrial robots. However, the principal rationale has always been, and is now, improved productivity.

Industrial robots offer a number of benefits which are the reasons for the rapid current and projected growth of industrial robot installations. These benefits are:

1. increased productivity;
2. improved product quality;
3. more consistent product quality;
4. reduced scrap and waste;
5. reduced reworking costs;
6. reduced raw goods inventory;
7. direct labor cost savings;
8. savings in related costs such as lighting, heating, and cooling;
9. savings in safety related costs; and
10. savings from correctly forecasting production schedules.

The Robot System

Work in a manufacturing setting is not accomplished by a robot. Rather, it is accomplished by a robot system. A robot system has four major components: the controller, the robot arm or manipulator, end-of-arm tools, and power sources (Figure 5-6). These components, coupled with the various other pieces of equipment and tools needed to perform the job for which a robot is programmed, are called the robot's work cell.

Figure 5-7 contains a schematic drawing of the work cell for Cincinnati Milacron's T3-726 robot, which is used for TIG welding. Notice that the work cell contains not just the robot system but also an index table with an operator's safety shield and special welding equipment.

The contents of a robot's work cell will vary according to the application of the robot. However, the one constant in a robot's work cell is the robot system, made up of a controller, robot arm, end-of-arm tools, and power sources.

The Controller

A robot is a special-purpose device similar to a computer. As such, it has all of the normal components of a computer, including the central processing unit, made up of a control section and an arithmetic/logic section, and a variety of input and output devices.

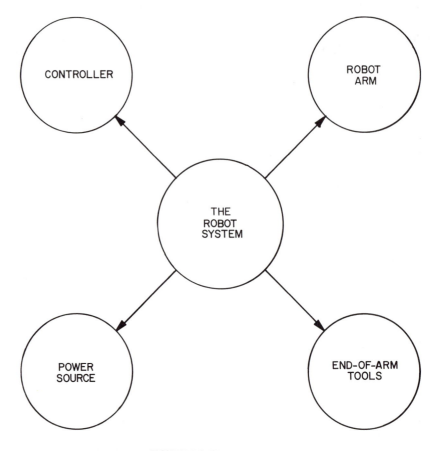

FIGURE 5-6 The robot system.

The controller for a robot system does not look like the micro-computer one is used to seeing on a desk. It must be packaged differently so as to be able to withstand the rigors of a manufacturing environment. Typical input/output devices used in conjunction with a robot controller include teach stations, teach pendants, a display terminal, a controller front panel, and a permanent storage device.

Teach terminals, teach pendants, or front panels are used for interacting with the robot system. These devices allow humans to turn the robot on, write programs, and key in commands to the robot system. Display terminals give operators a soft copy output source. The permanent storage device is a special device on which reusable programs can be stored. Figure 5-8 is a photograph of the Unimate Puma 200 robot system. In the background you will notice a display terminal with a special keyboard, an input/output module, and a teach pendant.

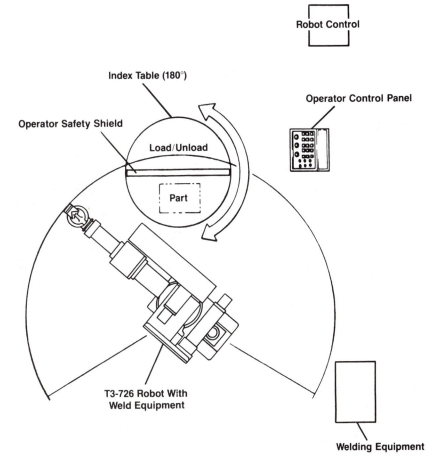

Robot Control

Index Table (180°)

Operator Control Panel

Operator Safety Shield

Load/Unload

Part

T3-726 Robot With
Weld Equipment

Welding Equipment

FIGURE 5-7 Work cell for the T3-726 industrial robot. Courtesy of Cincinnati Milacron.

Mechanical Arm

The mechanical arm, or manipulator, is the part of a robot system with which most people are familiar. It is the part that actually performs the principal movements in doing manufacturing-oriented work. Mechanical arms are classified according to the types of motions of which they are capable. The basic categories of motion for mechanical arms are rectangular, cylindrical, and spherical. These categories are dealt with in more depth later in this chapter. Figure 5-9 is a drawing of the mechanical arm for the Unimate Puma Series 700 robot.

End-of-Arm Tools

The human arm by itself can perform no work. It must have a hand, which in turn grips a tool. The mechanical arm of a robot, like the

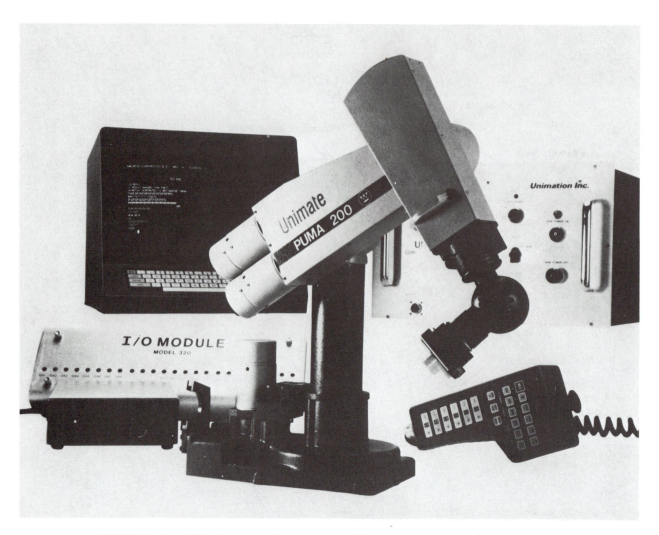

FIGURE 5-8 Puma 200 robot system. Courtesy of UNIMATION Incorporated, a Westinghouse Company.

human arm, can perform no work. It too must have a hand and a tool, or a device which combines both functions. On industrial robots the function of the wrist is performed by the tool plate. The tool plate is a special device to which end-of-arm tools are attached. The tools themselves vary according to the type of tasks the robot will perform. Any special device or tool attached to the tool plate to allow the robot to perform some specialized task is classified as an end-of-arm tool or an end effector.

The Power Source

A robot system can be powered by one of three different types of power. The power sources used in a robot system are electrical, pneumatic, or hydraulic. The controller, of course, is powered by electricity, as

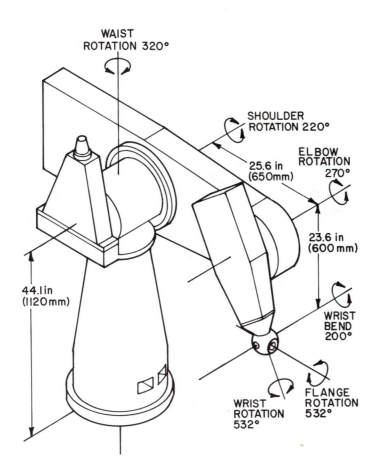

WAIST
ROTATION 320°

SHOULDER
ROTATION 220°

ELBOW
ROTATION
270°

25.6 in
(650mm)

23.6 in
(600 mm)

44.1 in
(1120 mm)

WRIST
BEND
200°

WRIST
ROTATION
532°

FLANGE
ROTATION
532°

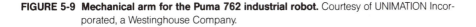

FIGURE 5-9 Mechanical arm for the Puma 762 industrial robot. Courtesy of UNIMATION Incorporated, a Westinghouse Company.

is any computer. The mechanical arm and end-of-arm tools may be powered by either pneumatic or hydraulic power. Some robot systems will use all three types of power. For example, a given robot might use electricity to power the controller, hydraulic power to manipulate the arm, and pneumatic power to manipulate the end-of-arm tool. Hydraulic power is fluid based. Pneumatic power comes from compressed gas.

Figure 5-10 shows a widely used industrial robot with a gripper device as an end-of-arm tool. The tool plate is the rectangular plate immediately behind the gripper. Each of the four components of a robot system will be examined in more depth in the sections which follow.

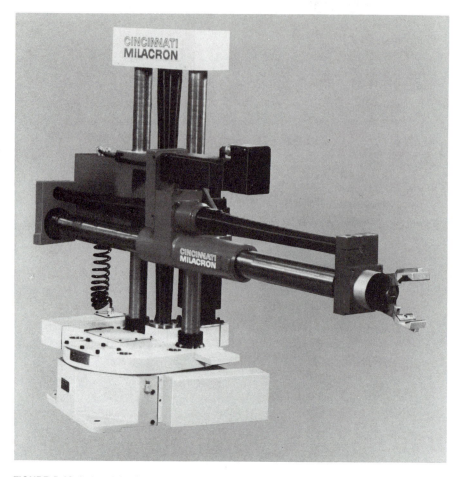

FIGURE 5-10 Industrial robot gripper. Courtesy of Cincinnati Milacron. (Note: Safety equipment may have been removed or opened to clearly illustrate products and must be in place prior to operation.)

Robot Terms and Phrases

There are a number of terms that are used repeatedly in the language of robotics. It is necessary to understand these terms in order to understand industrial robots. Some of the more frequently used terms are defined in this section and should be studied before proceeding to the remainder of this chapter.

Accuracy

Accuracy is a measure of how close a robot arm is able to come to the coordinates specified. There is always some difference between the actual and the desired point. The degree of difference is the accuracy of the robot.

Actuator

Any device in a robot system which converts electrical, hydraulic, or pneumatic energy into mechanical energy or motion.

Continuous Path

A servo-driven robot that provides absolute control along an entire path of arm motion, but with certain restrictions with regard to the degree of difficulty in changing the program.

Controlled Path

A servo-driven robot with a control system which specifies the location and orientation of all robot axes. A control-path robot moves in a straight line between programmed points.

Degrees of Freedom

The number of degrees of freedom of a robot is the number of movable axes on the robot's arm. A robot with four movable joints has four degrees of freedom.

End Effector

An end-of-arm tool which is attached to the robot's manipulator and actually performs the robot's work.

Fixture

A special device used to hold a workpiece in the proper position as it is being tooled.

Flexible Automation

An all-encompassing term which describes the flexibility, adaptability, and reprogrammable nature of modern industrial robots.

Limited Sequence

A simple, non-servo type of robot, sometimes called a "bang-bang" robot. Movement of a limited sequence robot is controlled by a series of stop switches.

Manipulator

Another name for the arm of the robot. It encompasses basic axes which control wrist movements for robots. The three basic axes are referred to as pitch, yaw, and roll.

Payload

The maximum weight a robot is able to carry at normal speeds.

Pitch

Up-and-down motion along an axis.

Point to Point

A robot with a control system for programming a series of points without regard to coordination of axes.

Repeatability

The degree to which a robot is able to return the tool center point repeatedly to the same position.

Roll

Circular motion along an axis.

Servo-Mechanism

An automatic feedback control system for mechanical motion.

Speed

The rate, in inches per second or millimeters per second, that the robot is able to move the tool center point.

Teach Pendant

A special control box which an operator uses to guide a robot through the motions required to perform a specific task.

Tool Center Point

A given point at the tool level around which the robot is programmed for performing specific tasks.

Work Envelope

The operating range, or reach capability, of a robot.

Yaw

Side-to-side motion along an axis.

Robot Applications

The list of manufacturing applications of industrial robots is almost unlimited. Robots are flexible machines. They are flexible in that they are reprogrammable and in that they can achieve all of the necessary motions required in manufacturing. They are also flexible in that they can handle a wide variety of manufacturing tasks. What follows is a list of some of the more common applications of industrial robots.

Arc welding
MIG welding
TIG welding
Palletizing
Stacking and unstacking
Loading and unloading
Drilling
Milling
Grinding
Deburring

Painting
Gluing
Parts handling
Movement of dangerous
   or toxic materials
Assembly
Loading and unloading
   of manufacturing machines
Cutting

Figure 5-11 is a schematic drawing of the T3-363 robot by Cincinnati Milacron used for unloading injection molding machines. Notice that the robot's work cell consists of an injection molding machine, a hot stamp machine, a bowl feeder, a pallet elevator, a conveyor, and, of course, the robot.

Figure 5-12 is a schematic drawing of the work cell for the T3-746 electric robot produced by Cincinnati Milacron. The application category of this robot is materials handling. Specifically, the robot is used for loading automobile dashboards into trim presses. This is an example of where a robot is used to perform work that could be dangerous for humans.

Figure 5-13, page 170, is a schematic drawing of the work cell for the T3-776 robot produced by Cincinnati Milacron. The application category for this robot is materials handling. Specifically, this robot is used to palletize unbound stacks of printed material. The robot picks up stacks of unbound printed material with a special forklift gripper. The stacks are placed on pallets. The pallets are automatically transported in and out of the robot's work cell. Separator tie sheets are placed on completed levels of palletized stacks by the robot.

Figure 5-14, page 171, is a schematic drawing of the T3-776 robot used, in this case, for machine loading. Specifically, this robot is used for loading and unloading machining centers.

Figure 5-15, page 172, is another schematic drawing of the T3-726 industrial robot produced by Cincinnati Milacron. The application category for this robot is cutting. Specifically, this robot is used for plasma arc cutting of three different shapes. These shapes are illustrated under "application cut-outs" in Figure 5-15. They are cut from the inside of a one-inch-thick, seven-foot-wide, ten-foot-high steel cylinder.

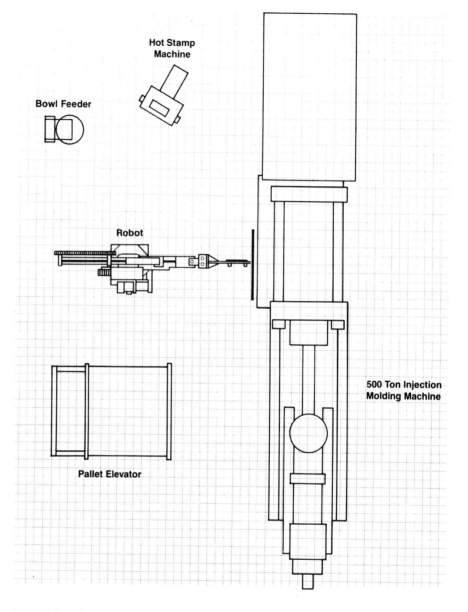

FIGURE 5-11 Schematic of the work cell of the T3-363 industrial robot. Courtesy of Cincinnati Milacron.

CATEGORIES OF ROBOTS

There are no industry standards which mandate a given approach to categorizing robots. In spite of this, it has become standard practice to categorize robots in one of six ways. These ways are by

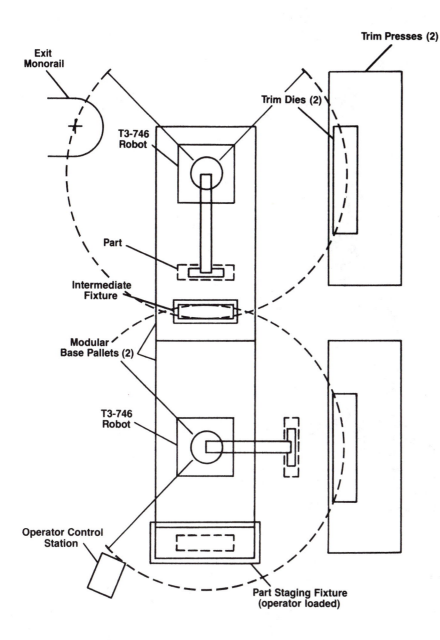

FIGURE 5-12 Schematic of the work cell of the T3-746 industrial robot. Courtesy of Cincinnati Milacron.

1. Arm geometry;
2. Power sources;
3. Applications;
4. Control technique;
5. Path control; and
6. Intelligence.

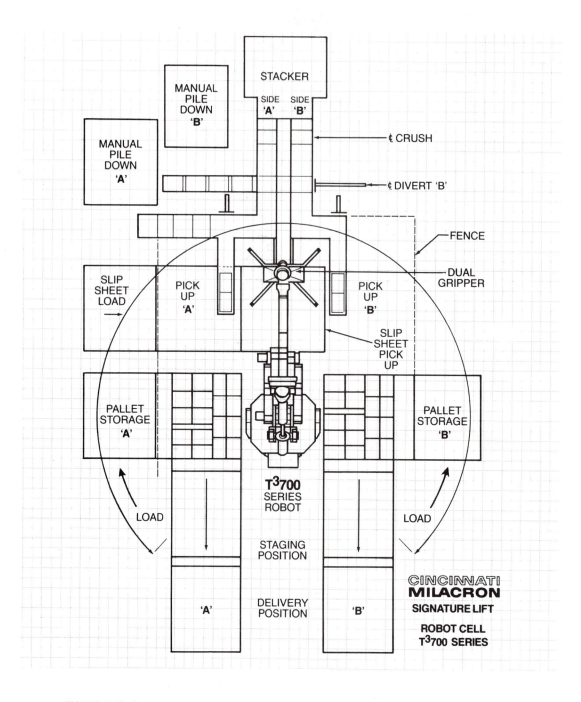

FIGURE 5-13 Schematic of the work cell of the T3-700 industrial robot. Courtesy of Cincinnati Milacron.

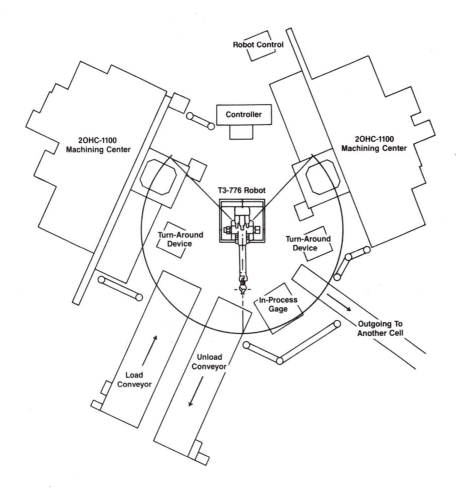

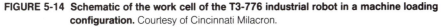

FIGURE 5-14 Schematic of the work cell of the T3-776 industrial robot in a machine loading configuration. Courtesy of Cincinnati Milacron.

Categorizing Robots by Arm Geometry

There are three classifications of arm geometry for industrial robots: rectangular, cylindrical, and spherical. A robot with rectangular arm geometry has a rectangular work envelope if viewed orthographically. If viewed in three dimensions, the work envelope is shaped like an elongated cube. An industrial robot with rectangular arm geometry is illustrated in Figure 5-16.

An industrial robot with cylindrical arm geometry has a work envelope which is described by a cylinder. Figure 5-17 is an example of an industrial robot with cylindrical arm geometry. An industrial robot with spherical arm geometry has a work envelope which is described by a portion of a sphere. Figure 5-18, page 174, is an example of an industrial robot with a spherical work envelope.

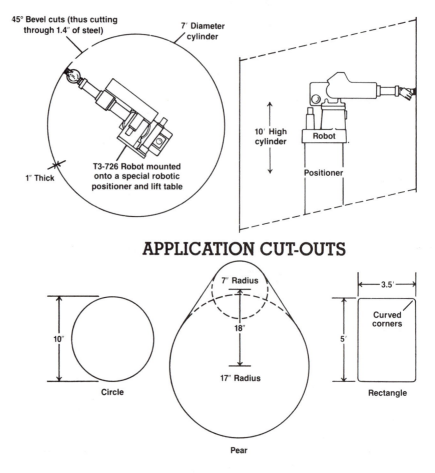

TOP VIEW

45° Bevel cuts (thus cutting through 1.4″ of steel)

7′ Diameter cylinder

T3-726 Robot mounted onto a special robotic positioner and lift table

1″ Thick

SIDE VIEW

10′ High cylinder

Robot

Positioner

APPLICATION CUT-OUTS

10″

Circle

7″ Radius

18″

17″ Radius

Pear

3.5′

Curved corners

5′

Rectangle

FIGURE 5-15 Schematic of the top and side views of the T3-726 industrial robot configured for cut-out applications. Courtesy of Cincinnati Milacron.

Each type of coordinate system has its individual characteristics as well as its advantages and disadvantages. A better understanding of industrial robots can be gained by closely analyzing the coordinate systems upon which their work envelopes are based.

Work Envelope Analysis

There are three types of coordinate systems associated with industrial robots: rectangular, cylindrical, and spherical. As the names imply, these systems make rectangular, cylindrical, and spherical work envelopes, respectively.

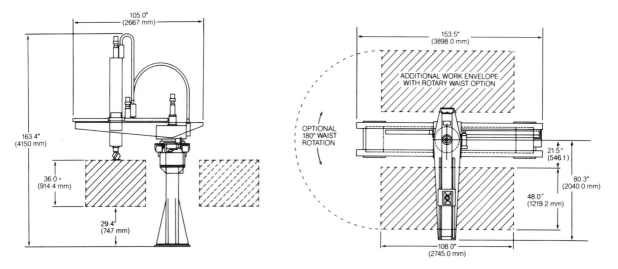

FIGURE 5-16 Schematic of the Unimate 6000 industrial robot. Courtesy of UNIMATION Incorporated, a Westinghouse Company.

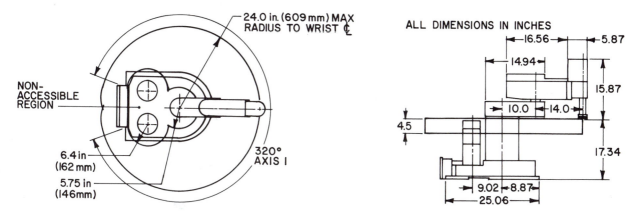

FIGURE 5-17 Schematic of an industrial robot with cylindrical arm geometry. Courtesy of UNIMATION Incorporated, a Westinghouse Company.

Figure 5-19 is an example of a rectangular work envelope. It is defined by imaginary lines running parallel to the X, Y, and Z axes. Figure 5-20 illustrates the X, Y, and Z movements of a rectilinear coordinate robot. Advantages of this type of geometry include (1) the potential for large work envelopes; (2) overhead mounting which frees up space on the shop floor; and (3) the allowance for simple, less sophisticated controllers. Disadvantages of this type of geometry are that overhead mounting structures (when used) can limit the access of materials handling equipment, and maintenance of overhead drive and control equipment is difficult because access is limited.

Figure 5-21 is an example of a cylindrical work envelope. It is defined by imaginary lines in the form of a vertical cylinder. Figure 5-22

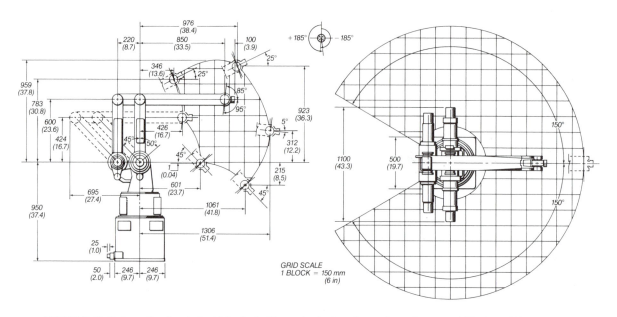

FIGURE 5-18 Schematic of an industrial robot with a spherical work envelope. Courtesy of Cincinnati Milacron.

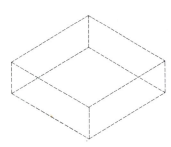

Figure 5-19 Rectangular work envelope.

illustrates the movements of a cylindrical coordinate robot. Advantages of this type of robot include: (1) deep horizontal reach into bins or machines; (2) requires relatively small amount of floor space; and (3) has the potential to handle heavy payloads. Cylindrical robots do have the disadvantage of limited left and right reach.

Figure 5-23 illustrates the movement of a spherical coordinate geometry robot. The work envelope of such a robot may have a variety of shapes. One such envelope is illustrated in Figure 5-18. Spherical geometry robots have the same advantages and disadvantages as cylindrical coordinate robots.

Categorizing Robots by Power Source

There are three power sources for industrial robots: electric, hydraulic, and pneumatic. All robots require electric power to operate the controller. The manipulator and end-of-arm tools may be operated by pneumatic or hydraulic power or a combination of both. A commmon configuration is one in which electrical power is used to operate the controller, hydraulic power is used to operate the manipulator, and pneumatic power is used to operate end-of-arm tools; but this varies from robot to robot.

Electrical Power

Electrical power is used to power the controllers in all industrial robot systems. However, electricity can also be used to power the manipulator and end-of-arm tools in applications with small payloads. Electric robots are able to perform tasks which require a high degree of accuracy and repeatability. Electrical power offers some advantages over pneumatic and hydraulic power. With electrical power, there is a comparatively low noise level and no problems with oil leaks and spills. Electrically driven robots are normally confined to applications involving payloads of 200 pounds or less.

FIGURE 5-20 Movements of a rectilinear coordinate robot.

Hydraulic Power

Hydraulic power is that which is provided by pressurized oil or some other type of fluid. Hydraulically powered robots have a number of disadvantages: They develop oil leaks, which sometimes cause oil spills in the work area; they are loud; and they can pose a fire hazard due to the oil leaks. However, in spite of these disadvantages, hydraulic power is able to achieve a very high power-to-size ratio. This means hydraulically powered robots can handle much larger payloads than electrically powered robots. For this reason, the hydraulically powered robot is the most frequently used in heavy-payload applications. In addition, hydraulically powered robots are more resilient than electrically powered robots. They have the ability to give when they make hard contact with an immovable surface.

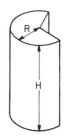

FIGURE 5-21 Cylindrical work envelope.

Pneumatic Power

Pneumatic power is that which is provided by pressurized gas. The configuration of a pneumatically powered robot is similar to that of a hydraulically powered robot, except that pressurized gas is used instead of pressurized oil. This solves the problem of oil leaks and oil spills in the working area. However, at present, the feedback control systems for pneumatically powered robots are not well developed. Consequently, pneumatically powered robots tend to be those which are referred to as "bang-bang" robots. Pneumatic power is much cleaner than hydraulic power, and it is better in high velocity applications where environmental considerations rule out the use of hydraulic power.

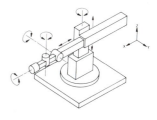

FIGURE 5-22 Movements of a cylindrical coordinate robot.

Categorizing Robots by Applications

Another way to categorize robots is by application. All of the various robot applications can be classified as either assembly or non-assembly applications. Applications such as welding, palletizing, stacking,

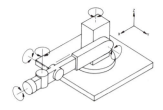

FIGURE 5-23 Movement of a spherical coordinate robot.

unstacking, loading, unloading, materials handling, drilling, milling, grinding, deburring, painting, and gluing are all non-assembly applications.

Assembly applications involve such tasks as soldering, press fitting, and applying threaded fasteners. Assembly applications typically require high accuracy and repeatability, but low payload capacities. For this reason, most robots used in assembly applications are electrically powered. Non-assembly robots are typically hydraulically and/or pneumatically powered because of the larger payloads involved. Figure 5-24 is a photograph of the Unimate Puma 700 series robot. The Puma 700 is an electrically powered robot used for assembly applications. Figure 5-25 is a photo of the T3-735 industrial robot produced by Cincinnati Milacron for such non-assembly applications as arc welding.

Categorizing Robots by Control Systems

When categorizing robots according to control systems, there are two broad categories of robots: servo and non-servo control robots. Servo-controlled systems are also referred to as closed-loop systems. Non-servo control systems are also referred to as open-loop systems. Servo system robots are more sophisticated and, in turn, more expensive. Non-servo system robots are simple and relatively inexpensive. In both cases, the control system refers to the method used to control the position of the robot tool.

Servo, or closed-loop, systems use a sensor device to continuously monitor the position of the tool as well as its direction and velocity along the desired path. Feedback from this continuous monitoring allows the position, direction, and velocity of the robot tool to be corrected continuously and as necessary to keep the tool on the desired path. Servo systems are used in those applications where path control is a critical element. Because of the feedback capability, servo or closed-loop robots are able to perform more complex manufacturing tasks and a wider variety of tasks.

Non-servo, or open-loop, robots do not employ sensors to continuously monitor the position, direction, and velocity of the robot tool. Rather, a series of physical limits or fixed stop points are used. Each joint in the robot must have these physical stops built in. The robot motion is from one extreme or stop point to the next extreme or stop point. There are no intermediate stops in an open-loop or non-servo robot. Non-servo robots are also referred to as "pick-and-place" or "bang-bang" robots.

These robots have the advantage of requiring less maintenance as well as being less expensive. However, they are limited to the more simple manufacturing tasks, require more complex end effectors, and tend to have a shorter life span than servo-controlled robots. Figure 5-26

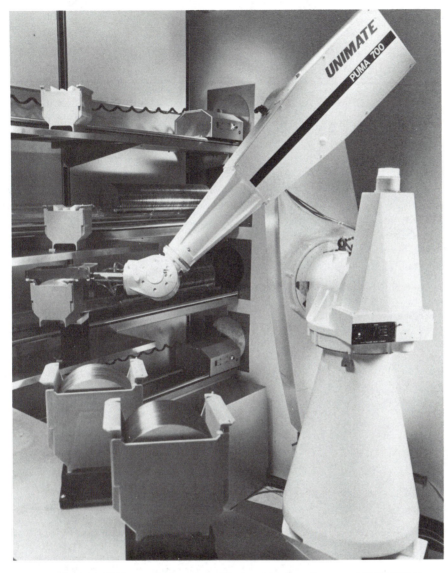

FIGURE 5-24 Puma 700 electrically powered industrial robot. Courtesy of UNIMATION Incorporated, a Westinghouse Company.

is an example of a servo or closed-loop control system robot. Figure 5-27 is an example of a non-servo or open-loop control system robot.

Categorizing Robots by the Type of Path Control

Path control refers to the method used by the robot controller to guide the end-of-arm tooling along the desired path. There are four types

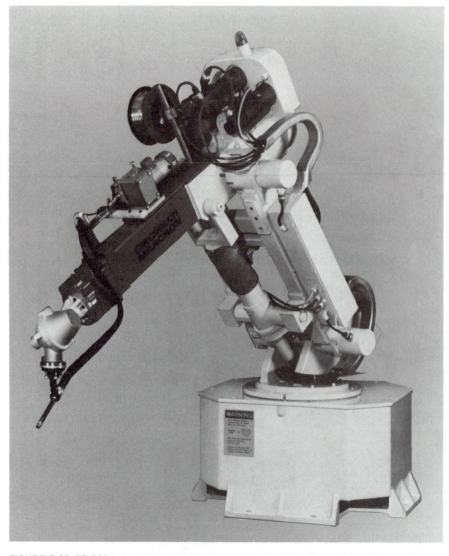

FIGURE 5-25 T3-735 industrial robot. Courtesy of Cincinnati Milacron. (Note: Safety equipment may have been removed or opened to clearly illustrate products and must be in place prior to operation.)

of path control used in modern industrial robots: stop-to-stop, continuous-path, point-to-point, and controlled-path.

Stop-to-stop path control is used in non-servo control system robots. This type of robot controls the path of the tool with a series of preset electronic or mechanical stop switches. These electronic or mechanical stops establish the extremes of motion for the robot and are responsible for the robot's ability to work its way through a sequence of motions required to complete a manufacturing task. Because of this

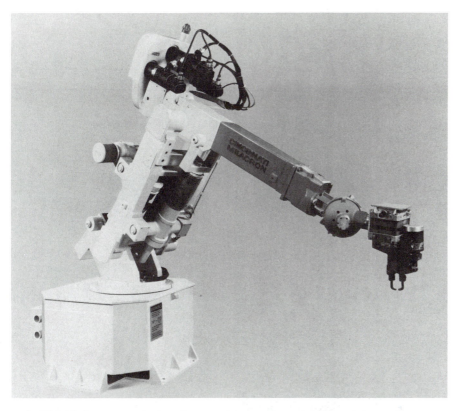

FIGURE 5-26 Servo-control robot. Courtesy of Cincinnati Milacron. (Note: Safety equipment may have been removed or opened to clearly illustrate products and must be in place prior to operation.)

simple, unsophisticated method of path control, stop-to-stop robots are generally used to perform only simple materials handling and line transfer tasks.

Point-to-point, continuous-path, and control-path systems are used with servo-controlled robots. Point-to-point path control is accomplished by programming a series of positions along a desired path of motion. Only stop points are specified. Point-to-point path control robots are used for materials handling and other applications where the control of movement between points is not a critical factor.

Continuous-path control robots are used in applications where the path of the tool is critical. With this type of control, every point along the path of motion is controlled. For this reason, the controller for a continuous path robot requires more memory than a point-to-point control robot. Every point along the desired path must be stored in the memory of the controller. Continuous-path control is generally not used in applications which involve frequent changes to the program.

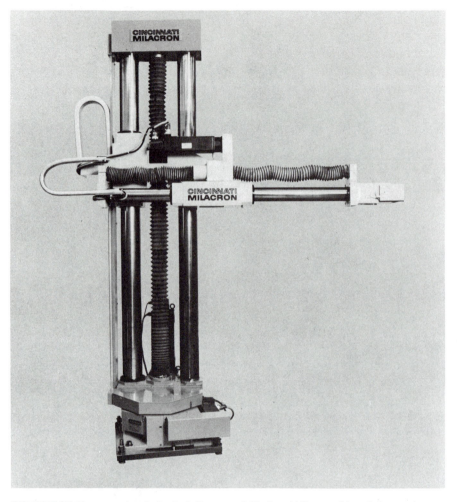

FIGURE 5-27 Non-servo control robot. Courtesy of Cincinnati Milacron. (Note: Safety equipment may have been removed or opened to clearly illustrate products and must be in place prior to operation.)

Controlled-path robots are those in which all axes along a path are coordinated. In this method, the path and the velocity at which the tool moves along the path are controlled. Controlled-path robots are typically used in processing, assembly, and welding applications.

Categorizing Robots by Levels of Intelligence

When robots are categorized according to their level of intelligence, there are three categories: high-technology, medium-technology, and low-technology robots. Those robots which are referred to as pick-

and-place or "bang-bang" robots are considered low-technology robots. High-technology robots are servo system robots which use either continuous-path or controlled-path control systems. It is more difficult to categorize robots in the intermediate group. These tend to be the more flexible non-servo robots and the more simple point-to-point servo system robots.

ROBOT END EFFECTORS

End effectors are the end-of-arm tools which actually do the work of the robot. The most frequently used end effectors come in two broad categories: grippers and special-purpose end effectors. Grippers, as the name implies, are used for gripping an object, performing the desired movement, and releasing the object. There are five different classifications of grippers: standard, vacuum, magnetic, air-pressure, and special-purpose (Figure 5-28).

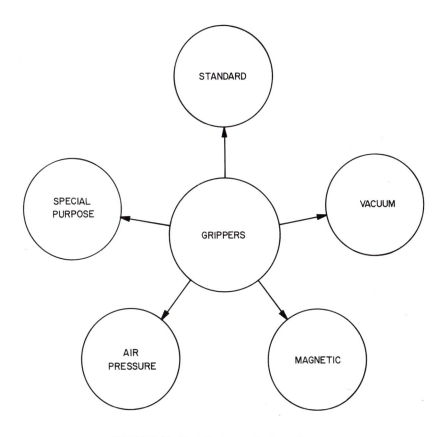

FIGURE 5-28 Classifications of robot grippers.

Standard grippers grasp the object between two fingers, somewhat like a mechanical claw. Figure 5-29 is an example of an industrial robot with standard grippers as end effectors. Vacuum end effectors use vacuum suction to pick up materials and magnetic grippers use magnetism. Air-pressure grippers use pneumatic fingers, which are particularly effective in applications requiring a gentle touch. Special-purpose grippers are those which are designed for applications where standard, vacuum, magnetic, and air-pressure grippers are not appropriate. Many special-purpose grippers are designed locally by the robot user.

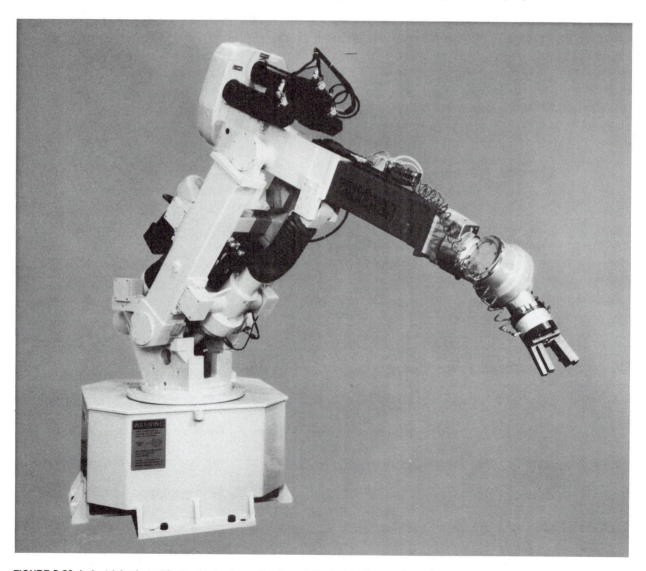

FIGURE 5-29 Industrial robot with standard gripper. Courtesy of Cincinnati Milacron. (Note: Safety equipment may have been removed or opened to clearly illustrate products and must be in place prior to operation.)

Special-purpose end effectors are not grippers. Rather, they are special-purpose tools designed to do specific jobs, such as drilling, welding, painting, grinding, and sanding (Figure 5-30). Figure 5-31 is an example of a robot with a special-purpose end effector for welding.

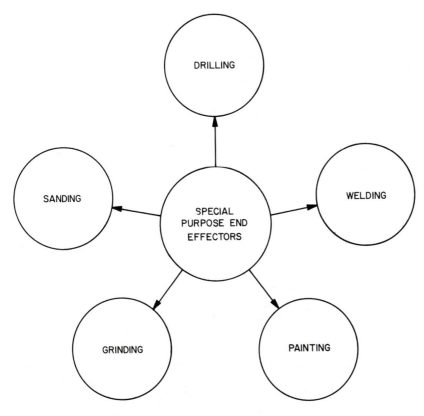

FIGURE 5-30 Types of special purpose end effectors.

ROBOT SENSORS

People are aware of their surroundings because they have five senses: sight, hearing, smell, taste, and touch. Using these senses, manufacturing personnel are able to perform the tasks required of them. For example, even the simplest task, such as picking up a tool, requires the senses of sight and touch.

Like people, robots must be aware of their environment if they are to perform manufacturing tasks. Sensors are devices which make robots aware. There are three basic types of robot sensors: contact, non-contact, and process-monitoring sensors (Figure 5-32). Before examining each type, the reader should become familiar with the purposes served by robot sensors.

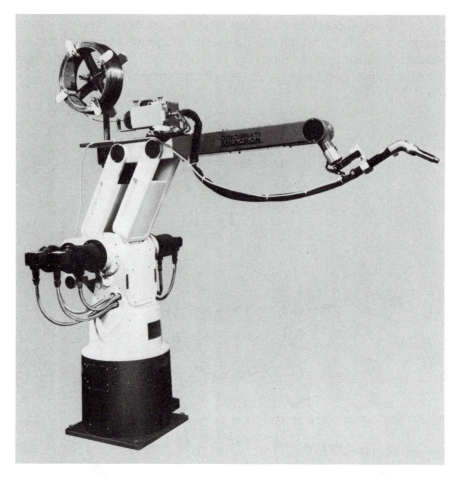

FIGURE 5-31 Industrial robot with a special-purpose end effector for welding. Courtesy of Cincinnati Milacron. (Note: Safety equipment may have been removed or opened to clearly illustrate products and must be in place prior to operation.)

Purposes of Robot Sensors

There are a variety of purposes served by robot sensors, all of which fall into one of three broad categories:

1. Monitoring
2. Detection
3. Analysis

In the first category, sensors allow robots to monitor the quality, consistency, and conformity of parts. They also allow robots to monitor factors that are important in assembly operations such as identifying the proper parts, ensuring that parts are properly located in the assembly, and ensuring that parts are properly oriented before putting them in the assembly.

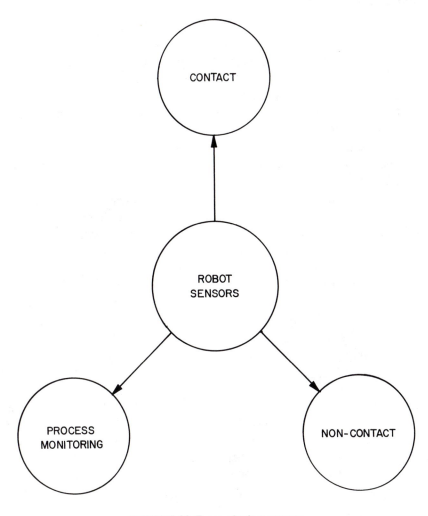

FIGURE 5-32 Types of robot sensors.

In the second category, sensors allow robots to detect circumstances that could be dangerous to the robot, the product, other equipment in the work cell, or to human workers. They also allow robots to detect system breakdowns or malfunctions.

In the final category, sensors allow robots to analyze parts for quality considerations. They also allow robots to analyze the robot system for internal problems so that the problems can be quickly and easily corrected.

Contact Sensors

Contact sensors are those which require the end-of-arm tool to actually touch the part. The most frequently used type of contact sensor is the limit switch. This is the least sophisticated type of sensor.

Limit switches are open-loop, non-servo devices used primarily with "bang-bang" or pick-and-place robots to detect motion and movement of parts within the work cell. There is no feedback channel between the end-of-arm tool and the controller.

Limit switches, like light switches, are electrical devices which are turned on and off mechanically by an actuator. The actuator is usually a part called a "dog."

A more sophisticated type of contact switch is artificial skin. Artificial skin is a closed-loop, servo-controlled gripper lining. A robot with artificial skin is able to vary its gripper pressure according to the physical properties of the part. This capability is particularly important for robots used in assembly operations.

Artificial skin is referred to as tactile sensing. Tactile sensing involves using a group of sensors arranged in a rectangular matrix or array. Each element in the matrix is able to sense a part and send feedback to the controller with regard to such factors as the part shape, texture, position, and orientation.

Non-Contact Sensors

Contact sensors, as you have just read, are similar to the human sense of touch in that contact must be made with the part in order for sensing to take place. Non-contact sensors, on the other hand, relate more closely to such senses as sight, hearing, or smell in that they do not require contact between the part and the robot. There are three principal types of non-contact sensors: proximity, photo-optic, and vision (Figure 5-33).

Proximity Sensors

Proximity sensors detect the presence of a part in a specified electromagnetic or electrostatic field. The actual range of the sensor's field varies from model to model and type to type. Proximity sensors come in a number of shapes. What shape works best in a given application depends on the shape of the part and the individual needs of the specific work cell.

There are proximity sensors available for sensing both metallic and nonmetallic parts. Metal-sensing proximity devices use a high-frequency electromagnetic field for sensing the part. Non-metal-sensing proximity devices use electrostatic capacitance as the sensing medium.

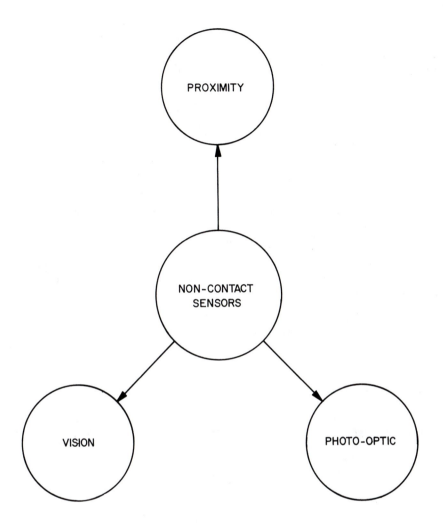

FIGURE 5-33 Types of non-contact sensors.

Photo-optic Sensors

Photo-optic sensors detect a part when it either breaks or reflects back a light beam. This technology is similar to that used in some types of burglar alarm systems. There are four principal types of photo-optic sensors available: separate, retro-reflective, diffuse-reflective, and definite-reflective sensors.

Separate sensor devices take their name from their configuration. This type of sensor system consists of two separate devices: one to produce and direct the light beam, and one to receive it. When the receiver fails to detect the presence of the light beam, it knows the beam has been broken.

Retro-reflective sensors are similar to separate sensors except that they use two light beams: one outgoing beam and one returning beam. Such a system has two hardware components: (1) a device to produce and send the outgoing light beam, as well as to receive the beam that is reflected back; and (2) a device for receiving the outgoing light beam and reflecting it back to the source. Sensing takes place when a part breaks either the outgoing or the reflected light beam.

Diffuse-reflective sensors bounce the light beam off of the actual part. Such a system has only one hardware component. It is a special device which produces and directs the light beam, then receives diffuse light reflected back from the part. A reflected beam of light is not necessary. Diffuse reflected light is sufficient for detection to occur.

Definite-reflective sensors are similar to both diffuse- and retro-reflective sensors. They are like diffuse-reflective sensors in that they use a single hardware device with both sending and receiving capabilities, and they bounce the light off of the actual object. They are like retro-reflective sensors in that what is reflected back must be a bona fide light beam; diffuse light is not sufficient to allow for detection.

Of the four types of photo-optic sensors, separate sensors have the longest range. In descending order, the other sensors are ranked retro-reflective, diffuse-reflective, and definite-reflective.

Vision Sensors

Vision sensors are the newest and most sophisticated of the non-contact sensors. Vision sensor systems consist of three components:

1. A camera and controller
2. Interface circuitry
3. Display terminal

The camera directed by the controller sees the silhouette of the part. The profile of the part is then displayed on the screen of the display terminal. The interface circuitry joins the camera and controller to the work cell and other systems interfaced with the work cell.

Vision systems allow robots to see what takes place in the work cell. Typical uses of vision systems include part orientation, inspection, location, and identification. Robot vision has enormous potential for expanding the capabilities and applications of industrial robots. However, it is a relatively new technology and much research is still needed. The current state-of-the-art is weak in such areas as recognition of 3-Dimensional images, selection of parts from non-ordered bins, or selection of random parts from within the same bin.

Process Monitoring Sensors

All of the sensors discussed so far have been robot sensors, which make a robot a more intelligent component of the work cell. Process monitoring sensors are not robot sensors. Rather, they are used for monitoring other manufacturing operations which interface with the work cell. When a problem arises in a manufacturing operation that is interfaced with the robot cell, process monitoring sensors warn the robot that corrective action is required.

ROBOT PROGRAMMING

Like computers, robots are capable of doing only what they are told. Also like computers, the communication link between robots and people is the program. In other words, people tell robots what to do through programs.

Robot programming languages have not yet been developed to the extent that computer languages have. There are not standards for robot programming languages at this point in time. Consequently, the tendency is for each vendor to develop its own language for its own robot. Standardization of robot programming languages is a high-priority goal of the robot industry.

Although robot languages are not yet standardized, they can be grouped according to the level of interaction they require of users. Using this approach, there are four classifications of robot languages (Figure 5-34):

1. Joint control languages
2. Primitive motion languages
3. Structured languages
4. Task-oriented languages

Joint Control Languages

Joint control languages are the lowest level of robot programming languages. They are used for controlling robot motion by controlling the positions of the manipulator in all axes for all joints. Joint positions are specified in angular form. Joint control languages are used with only the least sophisticated of robots. A major weakness of joint control languages is that they are written in terms of the motion of the robot manipulator at its various joints rather than in terms of solving the manufacturing problem.

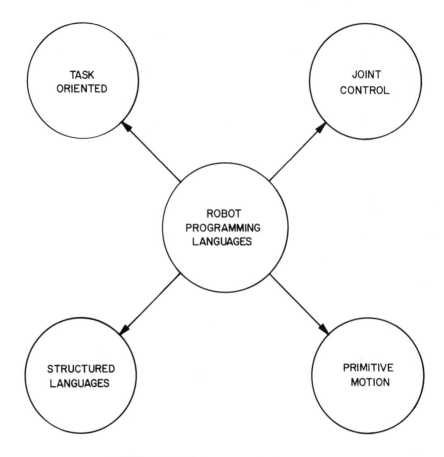

FIGURE 5-34 Robot programming languages.

Primitive Motion Languages

Primitive motion languages are the next higher level and the most widely used of the robot programming languages. Primitive motion languages are also called point-to-point languages. In order to write a point-to-point program, the robot manipulator must actually be moved to the various points which define the motion sequence. Each time a desired point is reached, a switch is activated making that point part of the program.

Although primitive motion languages are widely used, they have a number of disadvantages. The most glaring of these are that they cannot be written off-line and that they focus on the motion of the robot instead of on the task to be performed.

Structured Languages

Structured languages are the next higher level category of robot programming languages. The various languages which fall into this category begin to resemble computer programming languages. This is primarily due to their ability to use coordinate transformations and the fact that they allow for off-line programming.

Like computer programming languages, structured robot programming languages allow for branching and subroutines. They also resemble computer programming languages in that they require specialized training. This factor must be considered a negative. However, when weighed against the many advantages gained from structured languages—such as the use of transformations, the off-line programming capability, and branching and subroutine capabilities—the training factor becomes less of an issue.

Task-Oriented Languages

Task-oriented languages are the highest level of robot programming languages. These languages allow the user to actually program the robot in terms of the manufacturing problem rather than the motion of the manipulator. Programmers are able to use such normal conversational terms as "insert bolt" and "tighten nut." Because of this, task-oriented languages are easy and convenient for manufacturing people to learn.

However, task-oriented languages are not well-developed, nor are they widely available at present. Much research remains to be done at this level of robot programming. There are, however, several task-oriented languages currently in various stages of development. Robot productivity will reach new heights when task-oriented programming languages are fully developed and utilized.

ROBOTS AND PEOPLE

More than any other CAD/CAM-related technology, robots have the potential for controversy. The computer has extended human potential enormously, but what it does is classified as cognitive work, or in lay terms, "headwork." Robots, on the other hand, actually stand on the plant floor and perform tasks previously performed by human hands. In addition, in certain ways (the arm, wrist, etc.), they resemble human workers. These facts, coupled with the historical treatment of robots in science fiction, tend to make people wary of robots.

The original intent of robots was to perform work that was dangerous or work in which the human was little more than a robot himself. This latter type of work involved boring, repetitive tasks which people could not perform productively on a long-term basis. However, the uses of robots have already begun to extend beyond these limits.

There is no question that there are robots now doing work that otherwise would have been performed by humans. There is also no question that this is a trend that will continue. Consequently, there are people who have lost jobs to robots, and there will be people in the future who will lose jobs to robots. The theory is that the robots will create as many jobs as they will eliminate. This is a theory that is yet to be proved. In any case, robotics is a field with a great deal of potential for controversy.

As it is with the computer, the human element is still critical to the success of robots. In order for robots to be successful, they must be accepted by human employees. There must be people able to operate and program them and there must be people able to maintain and repair them. The key to success in making robots an effective part of a company's automation plan is education. There are three levels of education that are important with regard to robots (Figure 5-35):

1. That which is aimed at general awareness of all employees
2. That which is aimed at training operators and programmers
3. That which is aimed at training maintenance and repair technicians

General Awareness

Before installing robots in a plant, every employee at every level should be made aware of the ramifications. To limit awareness activities to manufacturing personnel can be a serious mistake. A shop worker who is worried about his job may carpool with an accountant, or socialize with an engineer, or play on a softball team with someone from contracts and bids. This employee and others like him or her will communicate their fears to their co-workers at all levels and in all functional areas of a given company. When one employee feels threatened, it can have a ripple effect that can harm the morale and productivity of all employees.

This issue is dealt with in greater depth in the chapter on CAD/CAM management. For now, it is important only to understand that awareness education for all employees is the critical first step in the successful implementation of robots.

Operation and Programming Training

Most operation and programming training is provided by vendors and relates specifically to the robot or robots in question. Important concerns in this area are:

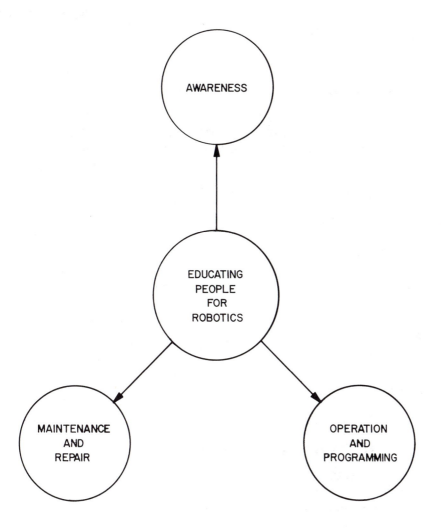

FIGURE 5-35 Levels of robotics education.

1. Operation and programming training should precede the actual installation of robots. This will involve off-site training at vendor-provided locations. It should be followed up with on-site training once the robots are installed.

2. If vendor training is robot-specific, which it normally is, additional training should be arranged through a local community college, technical school, college, or university that is non-specific. Operators and programmers will be more effective if they understand the broad picture, including the history of robots, the current state of robotics technology, advantages and limitations, common problems and solutions, and projected future developments. Ideally,

this broad conceptual learning should precede the more specific vendor-provided training.

Maintenance and Repair Training

A robot, like any other machine, will require maintenance and periodic repair. Training that is specifically geared toward a particular robot or robot system is provided by vendors. On-site repair of the more advanced robots is normally a matter of identifying a faulty printed circuit board and replacing it. Faulty boards, however, are not repaired on-site. Rather, they are returned to the appropriate vendor as a trade-in on new boards. The vendor then repairs and refurbishes the boards and resells them.

However, not all robots are of the sophisticated varieties, and there are a number of other problems that can occur which do not necessarily involve printed circuit boards. Consequently, beyond the vendor training provided, maintenance and repair personnel should pursue additional training in the four main areas of concern with regard to robot operation: mechanics, hydraulics, pneumatics, and electricity/electronics. Any maintenance or repair problems which occur with robots will fall into one or more of these areas. Maintenance workers who are well versed in these critical areas will be able to decrease the amount of downtime experienced by their company.

ROBOTICS-RELATED MATH

Robotics technicians frequently use algebra in performing such tasks as work power, energy, and forge calculations. The algebraic formulas most commonly used on the job in a robotics setting are summarized below.

$$W = D \times F$$

W = Work
D = Distance (In feet)
F = Force (In pounds)

Power is the amount of work accomplished in a specific period of time. The formula for calculating power is:

$$P = \frac{D \times F}{T}$$

P = Power (In foot-pounds per second)
D = Distance (In feet)
F = Force (In pounds)
T = Time (In seconds)

Energy is the capacity to do work. There are two types of energy: kinetic and potential. *Kinetic energy* is energy developed by a body in motion. The formula for calculating kinetic energy is:

KE = ½ X M X V²
KE = Kinetic Energy (In foot-pounds)
M = Mass of the Object
V = Velocity of the Object

Potential energy is energy that is possessed by a body at rest. The formula for calculating potential energy is:

PE = W X H
PE = Potential Energy (In foot-pounds)
W = Weight (In pounds)
D = Distance (In feet)

At times it is necessary to calculate mass and velocity in order to solve energy problems. The formulas for each are:

MASS

$$M = \frac{W}{32.16}$$

M = Mass
W = Weight (In pounds)
32.16 = Constant (Acceleration due to gravity
         in feet per second squared)

VELOCITY

V = 2 X g X D
V = Velocity
g = 32.16 Feet/second² (Constant)
D = Distance (In feet)

Torque is the effort (turning) required to rotate a mass or weight through a radius. The formula for calculating torque is:

T = F X D
T = Torque (Pounds/inch or foot)
F = Force (Weight of the object in pounds)
D = Distance (Of object from center of
         rotation in feet)

SUMMARY

A robot is a reprogrammable multifunctional manipulator designed to move material, parts, tools, or specialized devices through variable programmed motions for the performance of a variety of tasks. Developments which led to the current state of robotics began in the 1800s. However, robots which fit the definition above have been in existence for only twenty years.

The rapid growth of robotics has been caused by two factors: the development of the programmable integrated circuit, and economic conditions. A robot system has four main components: the controller, the robot arm or manipulator, end-of-arm tools, and the power sources.

Industrial robots have a number of different applications, all of which are classified as either assembly or non-assembly applications. Robots can be categorized in six different ways. These ways are according to: arm geometry, power sources, applications, control technique, path control, and intelligence.

Robots are made aware of their environment by sensors. There are three broad categories of robot sensors: monitoring, detection, and analysis. All sensors are either contact or non-contact sensors.

Robot programming languages are not standardized. Consequently, there are almost as many languages as there are robots. However, they can be grouped into four broad categories: joint control languages, primitive motion languages, structured languages, and task-oriented languages.

Robots have the potential to be controversial. Consequently, systematic education programs for all employees at all levels are recommended. Educational programs for employees of a company which plans to make robots part of its automation plan fall into three categories: (1) general awareness, (2) operator and programmer training, and (3) maintenance and repair training.

Chapter Five REVIEW

1. Define the term "industrial robot."
2. Who were (1) the producer and (2) the user of the first industrial robot used on an assembly line?
3. How many robots are expected to be in use in this country by 1990?
4. Explain how economics has contributed to the rapid growth of robotics.
5. List five benefits of industrial robots.
6. List and explain the four main components of a robot system.

7. Explain how robots can be classified according to arm geometry.
8. Explain the three power sources available to robots.
9. Explain servo and non-servo control systems and how they differ.
10. Explain low-, medium-, and high-technology robots and how they differ.
11. What are the two broad categories of robot end effectors? Give examples of each.
12. List and explain the various types of contact sensors.
13. List and explain the various types of non-contact sensors.
14. List and explain the four broad categories of robot programming languages.

One concept that must be in place before CIM can be realized is group technology. CIM systems are ideally suited for producing families of parts. Consequently, a company's product line must be arranged into families of parts in order for CIM to be implemented.

Major Topics Covered

- What Is Group Technology?
- Historical Background of Group Technology
- Part Families
- Parts Classification and Coding
- Advantages and Disadvantages of Group Technology

Chapter Six

Group Technology

The next chapter (Chapter 7) deals with computer integrated manufacturing (CIM). In order to understand this important CAD/CAM concept, it is first necessary to understand another: group technology.

WHAT IS GROUP TECHNOLOGY?

Group technology is a manufacturing concept in which similar parts are grouped together in parts groups or families. Parts may be alike in two ways:

1. In their design characteristics
2. In the manufacturing processes required to produce them

By grouping similar parts into families, manufacturing personnel can improve efficiency. Such improvements are the result of advantages gained in such areas as set-up time, standardization of processes, and scheduling.

Group technology can also improve the productivity of design personnel by decreasing the amount of work and time involved in designing new parts. Chances are a new part will be similar to an existing part in a given family. When this is the case, the new part can be developed by simply modifying the design of the already existing part. Design modifications tend to require less time and work than new design. This is especially true in the age of CAD. The advantages of group technology are dealt with at greater length later in this chapter.

HISTORICAL BACKGROUND OF GROUP TECHNOLOGY

Ever since the industrial revolution, manufacturing and engineering personnel have been searching for ways to optimize manufacturing processes. There have been numerous developments over the years since the industrial revolution mechanized production. Mass production and interchangeability of parts in the 1800s were major steps forward in optimizing manufacturing.

However, even with mass production and assembly lines, most manufacturing is done in small batches ranging from one workpiece to two or three thousand. In fact, even today over 70% of manufacturing involves batches of less than three thousand workpieces. Historically, less has been done to optimize small batch production than has been done for assembly line work.

There have been attempts to standardize a variety of design tasks and some work in queueing and sequencing in manufacturing, but until recent years design and manufacturing in small batch settings has been somewhat random.

The underlying problem that has historically prevented significant improvements to small batch manufacturing is that any solution must apply broadly to general production processes and principles rather than to a specific product. This is a difficult problem because the various workpieces in a small batch can be so random and different.

When manufacturing entered the age of automation and computerization, such developments as scheduling software, sequencing software, and materials requirements planning (MRP) systems became available to improve production of both small and large batches. Even these developments have not optimized the production of small batch manufacturing lots. The problem has become even more critical because, since the end of World War II, the trend has been toward more small batch and less production.

In recent years the problems of small batch production have finally begun to receive the attention necessary to bring about improvements. A major step is the ongoing development of group technology.

PART FAMILIES

It has already been stated that parts may be similar in design (size and shape) and/or in the manufacturing processes used to produce them. A group of such parts is called a part family. It is possible for parts in the same family to be very similar in design yet radically different in the area of production requirements. The opposite may also be true.

Figure 6-1 contains examples of two parts from the same family. These parts were placed in the same family based on design characteristics. They have exactly the same shape and size. However, you will notice that they differ in the area of production processes.

Part 1, after it is drilled, will go to a painting station for two coats of primer. Its dimensions must be held to a tolerance of plus or minus 0.125 inches. Part 2, after it is drilled, will go to a finishing station for sanding and buffing. Its dimensions require more restrictive tolerances of plus or minus 0.003 inches. The parts differ in material. The material for Part 1 is cold rolled steel; Part 2 is aluminum.

Figure 6-2 contains examples of two parts from the same family. Although the design characteristics of these two parts are drastically different (i.e., different sizes and shapes), a close examination will reveal that they are similar in the area of production processes. Part 1 is made of

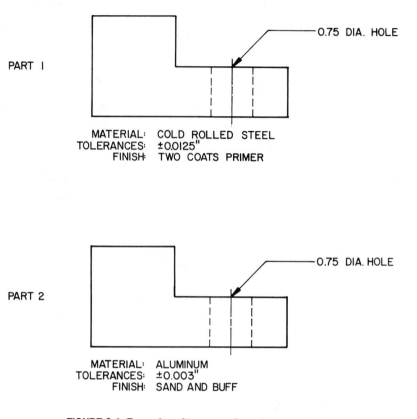

PART 1

0.75 DIA. HOLE

MATERIAL: COLD ROLLED STEEL
TOLERANCES: ±0.0125"
FINISH: TWO COATS PRIMER

PART 2

0.75 DIA. HOLE

MATERIAL: ALUMINUM
TOLERANCES: ±0.003"
FINISH: SAND AND BUFF

FIGURE 6-1 Examples of two parts from the same family.

stainless steel. Its dimensions must be held to a tolerance of plus or minus 0.002 inches, and three holes must be drilled through it. Part 2, in spite of the differences in its size and shape, has exactly the same manufacturing characteristics.

The parts shown in Figure 6-1, because of their similar design characteristics, were grouped in the same family. Such a family is referred to as a *design part family*. Those in Figure 6-2 were grouped together because of similar manufacturing characteristics. Such a family is referred to as a *manufacturing part family*. The characteristics used in classifying parts are referred to as "attributes."

By grouping parts into families, manufacturing personnel can cut down significantly on the amount of materials handled and movement wasted in producing them. This is because manufacturing machines can be correspondingly grouped into specialized work cells instead of the traditional arrangement of machines according to function (i.e., mills together, lathes together, drills together, etc.)

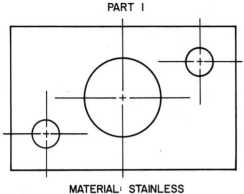

PART 1

MATERIAL: STAINLESS
TOLERANCES: ±0.002

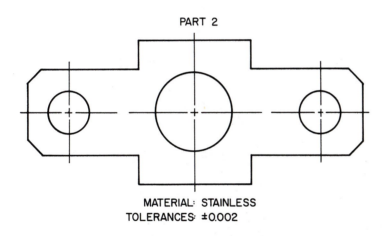

PART 2

MATERIAL: STAINLESS
TOLERANCES: ±0.002

FIGURE 6-2 Examples of two parts from the same family.

Each work cell can be specially configured to produce a given family of parts. When this is done, the number of setups, the amount of materials handling, the length of lead time, and the amount of in-process inventory are all reduced.

Grouping Parts into Families

Part grouping is not as simple a process as one might think. You already know the criteria used: design similarities and manufacturing similarities. But how does the actual grouping take place?

There are three methods which can be used for grouping parts into families (Figure 6-3):

1. Sight inspection
2. Route sheet inspection
3. Parts classification and coding

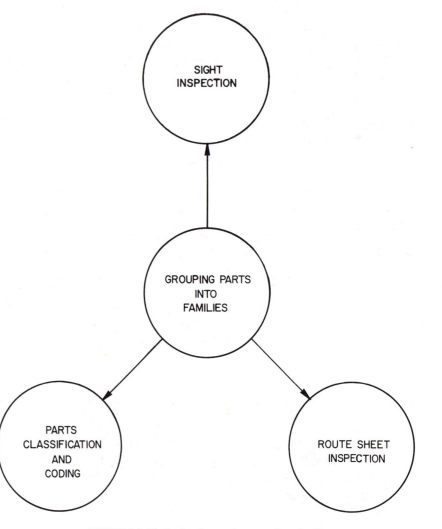

FIGURE 6-3 Methods of grouping parts into families.

All three methods require the expertise of experienced manufacturing personnel.

Sight inspection is the simplest, least sophisticated method. It involves looking at parts, photos of parts, or drawings of parts. Through such an examination, experienced personnel are able to identify similar characteristics and group the parts accordingly. This is the easiest approach, especially for grouping parts by design attributes, but it is also the least accurate of the three methods.

The second method involves inspecting the routing sheets used to route the parts through the various operations to be performed. This can be an effective way to group parts into manufacturing part families, provided the routing sheets are correct. If they are, this method is more accurate than the sight inspection approach. This method is sometimes referred to as the PFA or production flow analysis method.

The most widely used method for grouping parts is the third method: parts classification and coding. This is also the most sophisticated, most difficult, and most time-consuming method. Parts classification and coding is complex enough to require a more in-depth treatment than the other two methods.

PARTS CLASSIFICATION AND CODING

Parts classification and coding is a method in which the various design and/or manufacturing characteristics of a part are identified, listed, and assigned a code number. Recall that these characteristics are referred to as *attributes*. This is a general approach used in classifying and coding parts. There are many different systems which have been developed for actually carrying out the process, none of which has emerged as the standard.

The many different classification and coding systems which have been developed all fall into one of three groups (Figure 6-4):

1. Design attribute group
2. Manufacturing attribute group
3. Combined attribute group

Students of CAD/CAM should be familiar with how systems are grouped.

Design Attribute Group

Classification and coding systems which use design attributes as the qualifying criteria fall into this group. Commonly used design attributes include

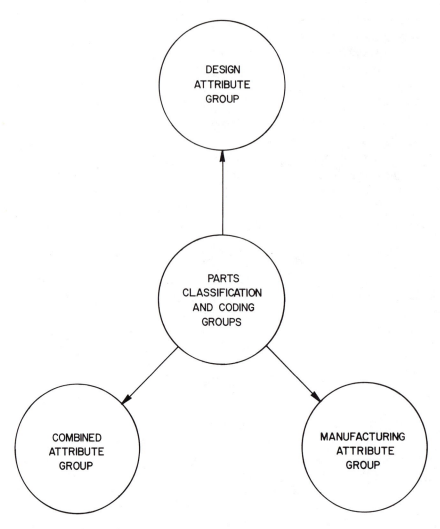

FIGURE 6-4 Parts classification and coding groups.

dimensions
tolerances
shape
finish
material

Manufacturing Attribute Group

Classification and coding systems which use manufacturing attributes as the qualifying criteria fall into this group. Commonly used manufacturing attributes include

production processes
operational sequence
production time
tools required
fixtures required
batch size

Combined Attribute Group

There are advantages in using design attributes and advantages in using manufacturing attributes. Systems which fall into the design attribute group are particularly advantageous if the goal is design retrieval. Those in the manufacturing group are better if the goal is any of a number of production-related functions. However, there is a need for systems which combine the best characteristics of both. Such systems use both design and manufacturing attributes.

Sample Parts Classification and Coding System

Some companies develop their own parts classification and coding system. But this can be an expensive and time-consuming approach. The more widely used approach is to purchase a commercially prepared system. There are several such systems available. However, the most widely used of these is the Opitz system. It is a good example of a parts classification and coding system and how one works.

Opitz System

Classification and coding systems use alphanumeric symbols to represent the various attributes of a part. One of the many classification and coding systems is the Opitz system. This system uses characters in 13 places to code the attributes of parts, and hence, to classify them. These digit places are represented as follows:

12345 6789 ABCD

The first five digits (12345) code the major design attributes of a part. The next four digits (6789) are for coding manufacturing-related attributes and are called the supplementary code. The letters (ABCD) code the production operation and sequence.

The alphanumeric characters shown above represent places. For example, the actual numeral used in each place can be 0-9. The numeral used in the "1" place indicates the length-to-diameter ratio of the part. The numeral used in the "2" place indicates the external shape of the part. The numeral used in the "3" place indicates the internal shape of the part. The numeral used in the "4" place indicates the type of surface machining. The numeral used in the "5" place indicates gear teeth and auxiliary holes. With such a system, a part might be coded as follows:

20801

The "2" means that the part has a certain length-to-diameter ratio. The first "0" means the part has no outstanding external shape elements. The "8" means the part has an internal thread. The second "0" means no surface machining is required. The "1" means the part is axial, not on pitch.

Figure 6-5 is a chart which shows the basic overall structure of the Opitz system for parts classification and coding. Figure 6-6 is a chart used for assigning an Opitz code to rotational parts. Figure 6-7 is a drawing of an actual part that has been assigned an Opitz code of 15400. This code was arrived at as follows:

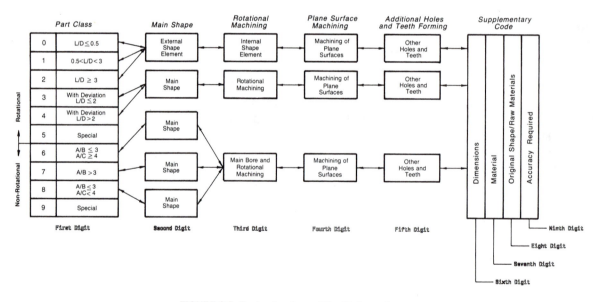

FIGURE 6-5 Basic structure of the Opitz system.

Step 1: The total length of the part is divided by the overall diameter:

$$\frac{1.75 \ (L)}{1.25 \ (D)} = 1.40$$

PART CLASS		MAIN SHAPE EXTERNAL		ROTATIONAL MACHINING INTERNAL		PLANE SURFACE MACHINING	ADDITIONAL HOLES AND TEETH
	L/D ≤ 0.5	Smooth, no shape elements		No hole, no breakthrough		No surface machining	No auxiliary hole
1	0.5 < L/D < 3	Stepped to one end or smooth	No shape elements	Smooth or stepped to one end	No shape elements	Surface plane and/or curved in one direction, external	Axial, not on pitch circle diameter
2	L/D ≥ 3		Thread		Thread	External plane surface related by graduation around a circle	Axial on pitch circle diameter
3	—		Functional groove		Functional groove	External groove and/or slot	Radial, not on pitch circle diameter
4	—	Stepped to both ends	No shape elements	Stepped to both ends	No shape elements	External spline (polygon)	No gear teeth — Axial and/or radial and/or other direction
5	—		Thread		Thread	External plane surface and/or slot, external spline	Axial and/or radial on PCD and/or other directions
6	—		Functional groove		Functional groove	Internal plane surface and/or slot	Spur gear teeth
7	—	Functional cone		Functional cone		Internal spline (ploygon)	With gear teeth — Bevel gear teeth
8	—	Operating thread		Operating thread		Internal and external polygon, groove and/or slot	Other gear teeth
9	—	All others		All others		All others	All others
First Digit		**Second Digit**		**Third Digit**		**Fourth Digit**	**Fifth Digit**

Rotational ↕ Non-Rotational

FIGURE 6-6 Assigning a code using the Opitz system.

Since 1.40 is greater than 0.5 but less than 3, the first digit in the code is "1."

Step 2: An overall description of the external shape of the part would read "...a rotational part that is stepped on both ends with one stepped end threaded." Consequently, the most appropriate second digit in the code is "5."

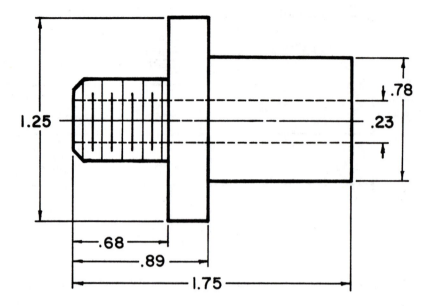

FIGURE 6-7 Sample part that can be coded using the Opitz system.

Step 3: A description of the internal shape of the part would read "...a through hole." Consequently, the third digit in the code is "4."

Step 4: By examining the part it can be seen that no surface machining is required. Therefore, the fourth digit in the code is "0."

Step 5: By examining the part it can be seen that no auxiliary holes or gear teeth are required. Therefore, the fifth digit in the code is also "0."

ADVANTAGES AND DISADVANTAGES OF GROUP TECHNOLOGY

There are several advantages which can be realized through the application of group technology. These advantages include

1. improved design,
2. enhanced standardization,
3. reduced materials handling,
4. simplified production scheduling, and
5. improved quality control.

Improved Design

Group technology allows designers to use their time more efficiently and productively by decreasing the amount of new design work required each time a part is to be designed. When a new part is needed, its various attributes can be listed. Then, an existing part with as many of these attributes as possible can be identified and retrieved. The only new design required is that which relates to attributes of the new part not contained in the existing part. Because this characteristic of group technology tends to promote design standardization, additional design benefits accrue.

Enhanced Standardization

Parts are classified into groups according to their similarities. The more similarities, the better. Consequently, enhanced standardization is promoted by group technology. As you have already seen, design factors account for part of the enhanced standardization. Setups and tooling also become more standardized since similar parts require similar setups and similar tooling.

Reduced Materials Handling

One of the major elements of group technology is the arrangement of machines into specialized work cells, each cell producing a given family of parts. Traditionally machines are arranged by function (i.e., lathes together in one group, mills together in another, etc.) A functional arrangement causes excessive movement of parts from machine to machine and back. Specialized work cells reduce such movement to a minimum and, in turn, reduce materials handling.

Simplified Production Scheduling

With machines grouped into specialized work cells and arrangement of parts into families, there is less work center scheduling required. There is also less scheduling within work cells. In addition to requiring less scheduling overall, group technology involves less sophisticated, less difficult scheduling.

Improved Quality Control

Quality control is improved through group technology because each work cell is responsible for a specific family of parts. This does two things, both of which tend to improve the quality of parts: (1) It promotes pride in the work among employees assigned to each work cell, and (2) it makes production errors and problems easier to trace to a specific source. In traditional manufacturing shops, each work center performs certain tasks, but no one center is responsible for an entire part. Such an approach does not promote quality control.

In spite of these advantages, there are several disadvantages associated with group technology. They have to do with transitioning from a traditional manufacturing approach to group technology. Rearranging the machines in a traditional shop into specialized work cells can cause a major disruption. Such disruptions translate quickly into monetary losses. As a consequence, some managers are reluctant to pursue group technology.

Another problem with group technology is that parts classification coding is a time-consuming, difficult, and expensive process, especially in the initial stages. This causes some managers to balk at the expense.

SUMMARY

Group technology is a manufacturing concept in which similar parts are grouped together in families. Such parts may be similar in either design or manufacturing characteristics. There are three methods used for grouping parts into families: (1) sight inspection, (2) routing sheet inspection, and (3) parts classification and coding.

In the parts classification and coding method, design and manufacturing characteristics, known as attributes, are assigned alphanumeric symbols. Each symbol indicates a specific attribute. There are many such systems, and no one has emerged as the standard. However, all fall into one of three groups: (1) the design attribute group, (2) the manufacturing attribute group, and (3) the combined attribute group.

Commonly applied design attributes include dimensions, tolerances, shape, finish, and material. Commonly used manufacturing attributes include production processes, operational sequence, production time, tools required, fixtures required, and batch size.

A frequently used system is the Opitz system which uses numbers (0-9) and letters (ABCD) in 13 places to code the various attributes, both design and manufacturing, of a part.

Group technology offers several advantages, including improved designs, enhanced standardization, reduced materials handling, simplified production scheduling, and improved quality control. However, the transition from a traditional manufacturing configuration to a group technology arrangement can be expensive and disruptive.

Chapter Six REVIEW

1. What is group technology?
2. How are parts grouped into families?
3. Explain the three methods used for grouping parts into families.
4. What is an attribute?
5. List five commonly applied design attributes.
6. List five commonly applied manufacturing attributes.
7. List and explain five advantages of group technology.
9. What are the disadvantages of group technology?

Thus far, each chapter has dealt with a specific and individual part of the overall CAD/CAM picture (i.e., the computer, CADD, CNC, and robotics). This chapter deals with what results when these various individual concepts or selected ones of them are integrated. The results are Computer Integrated Manufacturing (CIM) and Flexible Manufacturing Systems (FMS).

Major Topics Covered

- What are CIM and FMS?
- Rationale for CIM
- Resistance to CIM
- Types of CIM Systems
- Components of CIM Systems
- CIM System Applications
- CIM and People
- Projected Trends in CIM/FMS Technology

Chapter Seven

Computer Integrated Manufacturing (CIM) and Flexible Manufacturing Systems (FMS)

Comparisons to Chrysler's highly visible leader are deceptively easy. Both Lee Iacocca and Vittorio Ghidella have taken auto companies that were flirting with extinction and turned them around. Both have accomplished what few—including insider analysts—thought was possible. Both are aggressive, self-reliant, tough-minded.

But the comparison is too easy. The task faced by Ghidella posed many facets that were uniquely European, and the man himself, his background and experience, has proven matched to the task.

The Backdrop

When Ghidella took over at Fiat in 1978, the company was floundering: Cash poor, turning out few new models, facing unions that refused to permit layoffs, and with lifetime employment, many workers simply didn't work. Absenteeism was nearly 25%. Add to this that the Fiat labor force was infiltrated with terrorists who openly gunned down executives, one begins to sense the magnitude of Ghidella's task.

The first thing he did was address the problems of the shop floor. This was an arena he knew well, having spent time on plant floors in his early career with Fiat and then again at the Big Swedish bearing manufacturer SKF (Stockholm). He promptly identified and fired 60-some troublemakers above the protests of the union. Fully half were later proved to have terrorist affiliations. Then he laid off 20,000 workers, something that had never before been tried in Italy because of the powerful unions. The unions struck Fiat and held out for 35 days. In the end, however, 40,000 Fiat workers marched through the streets of Turin demanding that they be allowed to go back to work. Thus ended the strike, and with it the power of the Italian union. And once management held the choice to fire and to lay off workers, changes were swift in coming. Absenteeism fell to 5%. The number of cars produced per worker rose from less than 15 in 1979 to 30 in 1986. Interdepartmental conflicts fell from 4% to 1.5%, and conflicts between individual workers were completely eliminated.

Courtesy of "Manufacturing Engineering", May 1987. "Vittorio Ghidella: "Making Fiat Work" by Robin Bergstrom.

Vittorio Ghidella: Making Fiat Work

With order restored on the plant floor, Ghidella focused on other areas. An array of new Fiat models was planned, and at the same time a multibillion dollar investment program in capital improvements was launched. He ushered in new technology, robotics and flexible automation, reducing the break-even point from 1.6 million cars a year when he took over to 1.25 million. And he slashed operations that had become a financial drain—Fiat's sales operations in the US, for example.

Of course, not all these changes were met with warm embrace. A good many of Fiat's older managers balked—and a good many were replaced. But the result is that today Fiat is one of Europe's most efficient auto manufacturers—and one of its most profitable.

Role of Technology

Much of Ghidella's—and Fiat's—success is based upon innovative, resourceful management and the intelligent implementation of technology—on a multiplicity of levels. Fiat's operations are today heavily automated—from assembly and manufacturing operations to process monitoring and product design. "Technology," Ghidella says, "is a critical element in successful manufacturing—this for a reason often thought outside the process: the customer. The customer has become better educated and now requires things that were unnecessary only five years ago. I'm not referring to automotive gadgetry. The automobile today is a highly sophisticated, highly individualized common-use *tool*—it must operate day after day, reliably, yet must be highly stylized and sophisticated. And to deliver that tool, we have to employ the flexibility that the latest technological advances can provide."

Ghidella acknowledges that while the notion of flexible manufacturing is not new—people have, he says, been "talking" about it for years—today the opportunity to have flexibility has become a *necessity* if one is to survive the fierce winds of international competition. "Competition for

(continued)

customers in an already saturated market," he says, "is driving us to increasingly address a wider diversity of market niches. And as a company, we cannot make 1000 different models of cars. We must, however, offer a family of cars, and ideally we want to manufacture this family in a single installation. And that's what flexible manufacturing allows us to do—to produce a family of different products within the same installation utilizing the same processes and machinery."

This translates into practical experience in a number of ways. According to Ghidella, technology has radically altered the way we typically define manufacturing management. "Managing a manufacturing plant," he says, "once meant keeping track of materials, components, and processes. One had to check that the right materials were available, that they were processed properly, that the right manufacturing operations took place, that the quality was appropriate. To do this required a great deal of checking. One followed the production stream, one set priorities when something went wrong.

"Now, however, computers allow us to track and to check any phase of the manufacturing process at any given time. Further, many of the basic processes themselves are entirely automated. The modern refinery is closely analogous to the modern manufacturing plant. In the refinery, computers control all of the processes—start or stop them, speed them up or slow them down, make adjustments. The computer is the supervisor."

Ghidella does not, however, suggest that the manufacturing engineer or manager is becoming obsolete. On the contrary, he sees the role of the manufacturing manager remaining critically important, if somewhat different in content. "Manufacturing managers," he says, "will no longer check or monitor processes. That is a thing of the past. In today's modern manufacturing plants—and certainly in tomorrow's—manufacturing managers will act, make decisions, based on information fed by computers monitoring controls, which will be checking the actual processes."

Role of the Worker

At many of today's upscale manufacturing operations, technology has had an impact on job content and execution. Fiat is no exception, especially at the ends of the process spectrum—the design function, on the one hand, and the manufacturing function, on the other.

Ghidella is committed to the full integration of design and manufacturing through the effective implementation of CAD/CAM. And Fiat has proven this merger to be successful in the design and manufacture of its popular Uno car model. "The advantages of integration through CAD/CAM," Ghidella says, "are tremendous. We constantly improve our processes. We make fewer mistakes, have less idle time in the plant, speed up model innovation—the list goes on."

However, the computer is not a panacea. Ghidella: "Make no mistake about it, a computer does not design a car. This is still the role of a human being. It takes a *person* to design—whether it's the car, a wheel, or a fender. But once this step is complete, full computerization takes over—CAD/CAM for design checks, simulation, machining, and so on. But the human being as designer cannot be replaced."

If the role of the designer seems above scrutiny, the same is not so on the manufacturing floor. The ranks here, through the implementation of automation and flexible technology, have been thinned. In fact, Ghidella predicts the demise of certain jobs altogether. "For example," he says, "the typical assembly line worker, I feel, will disappear. Maybe not today, but certainly in the future. And this is a delicate matter, for the human element in manufacturing is critical, is the key element to the success of any system. The crucial point is knowing how best to integrate the people you need into the modern manufacturing processes."

This integration, according to Ghidella, can best be addressed through the proper application of training. He indicates that an entirely different menu of skills is needed by the workers in modern manufacturing facilities. "Computerization and automation have given the worker an entirely different role," Ghidella says. "Gone, by and large, is the need for

(continued)

superior physical strength on the plant floor. Today the worker must have responsibility, company loyalty, adaptability. He or she must know about machine controls, utilization, computers, robots. These new workers must be more than mere workers—there must be an added dimension to their role."

Single Decision-Maker

Although it's evident that Ghidella sees new roles for plant workers in the future, he's less enthusiastic about some of the more experimental roles and relationships being played out in many of today's manufacturing operations. The team approach to manufacturing is one such experiment. "Sharing responsibility is one thing," he says, "but asking a group to issue a common decision is entirely another."

His point is simple: The boss is still—regardless of all else—the boss. "It's the responsibility of the top person," he explains, "to make decisions. It's the boss's responsibility as well to listen to others, to take advice, to try to understand through others what the situation is. But in the end, the boss must decide."

He describes how things work at Fiat. He refers to the Fiat organization as a functional one, with nine individuals reporting directly to him. These individuals are responsible for particular functional areas—manufacturing or marketing, for example. "We make a group of ten," he says, "and we meet to discuss problems, plans, and so on. But I must make the final decision. It's not a risk-free process, but it's one that works. The other, trying to make collective decisions, simply does not work."

Another notion about which Ghidella is a little cautious is the quality circle. Fiat uses a version of these circles but somewhat differently. "We use them," says Ghidella, "to show the worker why a car is designed the way it is and why quality is important. But workers can't improve quality just by belonging to a quality circle. Quality must be resolved at the beginning—during design or manufacturing—never at the end of the process."

The Importance of Perspective

"I try very hard not to be too much of an engineer," Ghidella says. It's a remark that strikes an odd note but one that makes sense. By trying not to be too much of an engineer, Ghidella is trying to keep his perspective. "If somebody spends too much time getting one kind of experience—any kind of experience—it affects one's judgment. One cannot be objective. One's judgment will not be clear."

He suggests that regardless how bright and intelligent one may be, if a diversity of human interaction is missing from one's work, human understanding will likely be missing as well. "And that's crucial," Ghidella says, "because you have to care about all the elements, the components, the aspects. To work in manufacturing has been important to me because not only did I have to work with the machinery, to understand the machinery, but I also had to work with—and *understand*—the people. And, that's important."

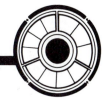

WHAT ARE CIM AND FMS?

Computer integrated manufacturing is the most modern, most automated form of production. It involves tying different phases of production together into one wholly integrated system. The term "flexible manufacturing system" is sometimes used synonomously with computer integrated manufacturing system. Actually, however, a flexible manufacturing system is one type of CIM system designed for medium range production volumes and moderate flexibility. Other types of CIM systems are special systems and manufacturing cells. There are other terms which have been used to describe the same concept, but CIM and FMS are the two which are most widely used today.

Although people have come to think of CIM as an ultramodern concept, it is really a very old concept in new form. The earliest approach to manufacturing was a wholly integrated approach. In the earliest days of manufacturing, one craftsman performed all of the various tasks related to producing a manufactured product. Because the craftsman performed all of the various tasks, the tasks were integrated. The tool used to accomplish this total integration of production processes was the craftsman's mind.

Picture an early craftsman cutting down trees, stripping them of their bark, splitting them into planks, cutting them into shapes, and forming them into wooden products such as furniture or even shelter. This was human integrated manufacturing. Then, in the late 1700s and early 1800s, technological advances led to specialization in manufacturing. In the era of specialization, one person was assigned one task to accomplish. That person accomplished the same task over and over. Specialization led to the demise of integration. Specialization led to such concepts as production control and quality control. They were necessary because each person or groups of people involved in the process only accomplished that small specialized task to which he was assigned. Consequently, someone had to oversee the entire operation to ensure that the finished product was what it was meant to be.

Specialization remains the norm in manufacturing. However, in the mid-1980s a new concept began to be seen more and more. That concept was known variably as Computer Integrated Manufacturing or Flexible Manufacturing Systems. This modern-day technological innovation is really nothing more than a new version of the total integration which existed in the days of the craftsman. This time, however, it is the computer rather than the mind of the craftsman that allows for total integration.

There are many different types of computer integrated manufacturing systems. This is because each system is designed to meet the specific needs of the individual manufacturing setting where it will be used. However, in general, a CIM system is any computerized manufacturing system in which numerically controlled machines are joined together and connected by some form of automated material handling system (Figure 7-1).

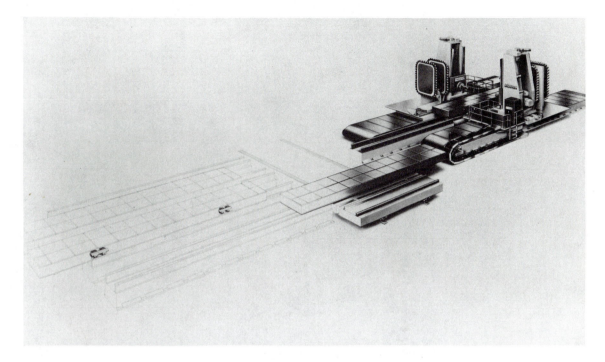

FIGURE 7-1 Example of a CIM system. Courtesy of Cincinnati Milacron.

This figure shows Cincinnati Milacron's XL series traveling column CNC machining centers linked together in an automated cell. In this system, two CNC machining centers are linked together by a cart-type material handling system. The cart system is further linked to four part load/unload stations to permit untended operation.

In a CIM system such as that shown in Figure 7-1, the computer is used in several ways. (1) It is used to control CNC machines. (2) It is used to control the materials handling system. (3) It is used for monitoring all production tasks accomplished by the system.

Figure 7-2 is a photograph of a flexible manufacturing machine/robot cell produced by Cincinnati Milacron. In this CIM system, rough castings are machined and inspected. The system contains two CNC machines, gaging devices, a conveyor, and two robots. The rough castings are loaded onto the conveyor by a robot and finished castings are unloaded by a robot.

Figure 7-3 is a photograph of the ACRAMATIC 975C composite tape laying (CTL) system produced by Cincinnati Milacron. The CTL

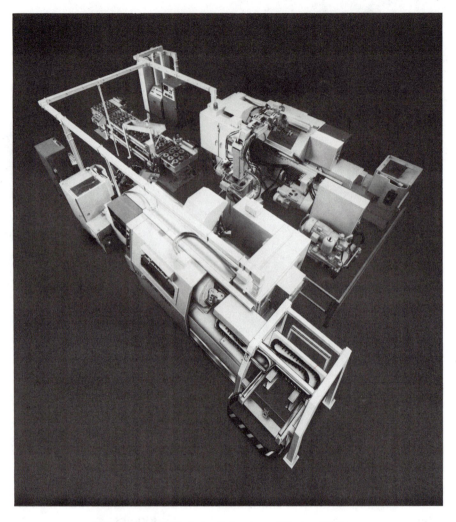

FIGURE 7-2 FMS/robot cell. Courtesy of Cincinnati Milacron.

system automatically dispenses tape at speeds of 100 feet per minute. Tape can be automatically laid in 3-inch, 6-inch, or 12-inch widths, debulked, cut, and overlaid ply-on-ply. Such systems are used in the manufacture of aircraft and space products.

In spite of advances in manufacturing automation, there is human involvement with a CIM system. Typically, this involvement falls into four broad categories: (1) loading raw stock and materials onto the system for processing, (2) unloading processed workpieces from the system, (3) changing tools on machines within the system, (4) setting tools on machines within the system, (5) continuous maintenance of the system, and (6) occasional repair of the system when there is a breakdown or malfunction.

FIGURE 7-3 ACRAMATIC 975C CTL system. Courtesy of Cincinnati Milacron.

RATIONALE FOR CIM

Stand-alone CNC machines are used in low-volume manufacturing applications which require a high degree of flexibility in order to produce a wide variety of parts. They represent one extreme on the manufacturing spectrum. At the other extreme are transfer lines. Transfer lines are used in high-volume manufacturing applications where all parts produced are identical. Transfer lines are not flexible; they cannot accommodate variety.

Transfer lines are a major component of what is sometimes referred to as "Detroit Automation." A transfer line consists of several workstations linked together by materials handling devices that transfer workpieces from station to station. The first station holds the raw material; the last station is a bin for collecting the finished parts; each station in between performs some type of operation on the parts as they pass through. The transfer of parts from station to station and the work performed on them are automatic. Transfer lines are appropriate in situations which involve high-volume production of identical parts.

There has always been a gap between these two extremes and there has always been a need to fill this gap. The medium-volume, moderate-flexibility manufacturing situation has long been a problem in need of a solution. CIM systems are such a solution. CIM fills the void between stand-alone CNC machines and transfer lines (Figure 7-4).

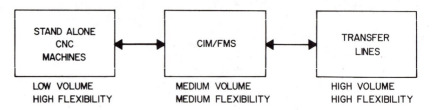

STAND ALONE CNC MACHINES	CIM/FMS	TRANSFER LINES
LOW VOLUME HIGH FLEXIBILITY	MEDIUM VOLUME MEDIUM FLEXIBILITY	HIGH VOLUME HIGH FLEXIBILITY

FIGURE 7-4 The manufacturing continuum.

CIM offers a number of advantages that, when taken together, form the rationale for this modern approach to medium-volume, flexible production. The most important of these are:

1. Produces families of parts
2. Accommodates the random introduction of parts
3. Requires less lead time
4. Allows a closer relationship between parts to be produced and workpieces loaded onto the system
5. Allows better machine utilization
6. Requires less labor

Families of Parts

A family of parts is a group of parts in which individual members are similar either in size and shape or in the manufacturing processes they must undergo during production. This concept was covered in Chapter 6. The inherent flexibility of CIM systems, which can be attributed to the reprogramability of the CNC component, allows them to produce families of parts.

Random Introduction of Parts

The manufacturing requirements of each individual workpiece in a family of parts are stored in the computer control component of the CIM system. This allows random introduction of parts onto the system. As an individual part is loaded onto a CIM system, its identification is fed to the computer controller.

The computer, in turn, transfers it to the appropriate machine(s) within the system. These machines have already been appropriately programmed and tooled. This capability is a strength of CIM systems, but it can also be a weakness without wise production planning.

Because each machine in the CIM system has been tooled and programmed to perform certain tasks, it is necessary to plan production runs carefully in order to ensure maximum machine utilization. If a large number of the same part are introduced at the same time, one machine in the system may be used extensively while others go unused. Avoiding such inefficient use of the system requires careful planning and the concurrent introduction of the optimum variety of parts.

Less Lead Time

CIM systems combine in one integrated system many of the separate machines of the traditional manufacturing shop. This and several other factors reduce the amount of lead time associated with CIM. The most important of these are: (1) less time spent in tooling setup, and (2) less time spent in workpiece setup.

Tooling setup time is reduced with CIM because the tooling takes place off-line. Workpiece setup time is reduced because it too is accomplished off-line on pallets. A given workstation can be programmed to handle several pallets, each carrying a different workpiece from among the family of parts.

Parts Produced vs. Workpieces Loaded

The number of workpieces loaded onto a CIM system relates more directly to the number of parts to be produced. Because the amount of lead time is significantly less and because too many workpieces can clog the system, the in-process inventory for CIM systems is less than that of traditional batch processing systems.

Better Machine Utilization

CIM allows for significantly better machine utilization if the proper production planning has been accomplished. Various factors lead to better machine utilization with CIM. These include less tooling setup time, less workpiece setup time, and the ability to produce families of parts.

Less Labor

CIM systems reduce the amount of direct and indirect labor costs associated with finished workpieces. This is because CIM systems require less human involvement in producing them. Ten to twelve traditional CNC machines require ten to twelve operators. A CIM system with ten to twelve CNC machines might require as few as four people. This means less direct labor. Indirect labor costs resulting from such tasks as materials handling are also reduced with CIM. This is because most material handling in CIM systems is automated.

RESISTANCE TO CIM

In spite of its potential, there is resistance to CIM at both the management and worker levels. Much of the resistance can be attributed to a lack of understanding. CIM is not a concept that has come about as a result of a push from management. The potential of CIM was first realized by skilled workers on the shop floor.

Forward-looking shop floor personnel have long felt the need for better integration of the many specialized processes and tasks involved in the production of manufactured goods. This is only natural coming from descendants of craftsmen who, in their time, achieved total integration by performing all of the various tasks required to produce manufactured goods. These innovative shop personnel saw the computer as a way to once again achieve total integration.

However, as is the case with any new concept, its potential in such areas as productivity improvement, direct and indirect labor cost savings, and machine utilization must be projected and estimated. Managers who are asked to consider a substantial initial investment are apt to want hard actual figures rather than estimates or projections. Of course, the only way to obtain such figures is to implement the concept and evaluate.

This dilemma has inhibited the acceptance of CIM. Further inhibiting the acceptance is the distance which sometimes exists between those who understand and want CIM and those who make the decisions. Shop personnel speak of the need to integrate such processes as materials handling, machining, and assembly. Managers speak in terms of initial investment costs, hidden costs, return on investments, and depreciation schedules. Such disparities have inhibited the acceptance of CIM.

In spite of these problems, CIM has begun to achieve a foothold in the world of manufacturing. Each time an innovative manager decides to undertake the implementation of CIM, hard data become available to other managers concerning such factors as productivity improvements, hidden costs, return on investments, and so on. In addition, vendors of CIM systems have taken the lead in bridging the language barriers between management and shop personnel.

In a way, these vendors act as interpreters by converting the concerns of shop personnel into terms understood by managers, and vice versa. Such efforts are having a positive effect. Leading manufacturers such as General Motors are making strides toward their goal of totally integrated, totally automated factories. However, progress is slow, and there are many problems.

One of the most difficult of these is resistance from shop workers. Not all shop workers accept CIM. To many, it represents the threat of occupational obsolescence. The goal of the fully automated factory is not a goal shared by workers who will lose their jobs as a result of it. Worker resistance to CIM is of the same nature as that experienced with such other CAD/CAM technologies as CAD, CNC, and robotics.

Worker resistance is one more factor which mitigates against management's acceptance of CIM, since responsible managers will feel compelled to make retraining and job placement assistance available to employees whose jobs are eliminated by CIM. They will also be compelled to add all the costs of such services to the projected costs of implementing CIM.

TYPES OF CIM SYSTEMS

CIM systems fall between transfer lines and stand-alone CNC machines on a graph of production volume versus part variety flexibility. By applying these same criteria—production volume and part variety—CIM systems can be divided into three categories (Figure 7-5):

1. Special systems
2. Flexible manufacturing systems
3. Manufacturing cells

The systems above are listed in order from the least flexible (special systems) to the most flexible manufacturing. Of all CIM systems, those classified as special systems are capable of the most volume production and the least flexibility. Special systems might produce as many as 15,000 parts per year, but they are limited to less than ten different part types.

Flexible manufacturing systems represent the middle group in CIM systems. An FMS might have a production volume as high as 2,000 parts per year and a capability of handling as many as 100 different parts.

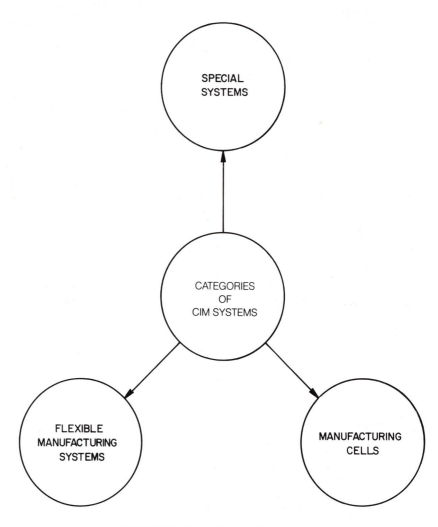

FIGURE 7-5 Categories of CIM systems.

The most flexible of CIM systems are manufacturing cells which, in turn, are capable of the lowest production volumes. A manufacturing cell might have a production volume as low as 500 parts per year, but a capability of handling as many as 500 different part types.

COMPONENTS OF CIM SYSTEMS

CIM systems vary in configuration, but regardless of the configuration, they all have the same three principal components:

1. Computer control component
2. Machine tool component
3. Materials handling component

Computer Control Component

The computer control component is the brains of the CIM system. It uses stored data files to control all aspects of system operation. There are several different types of data files associated with CIM systems:

1. Part data file
2. Routing data file
3. Part production data file
4. Pallet data file
5. Workstation data file
6. Tool-use data file

The part data file is where programmed instructions for each part are stored. The routing data file contains the route each part will take through the various workstations in the system. The part production data file contains all of the production data for each part, including such information as inspections required and production rates. The pallet data file contains information identifying each pallet and the parts which go with it. The workstation data file contains the control codes for the tools used at each workstation. The tool-use data file contains the tool life for all tools in the system. It also holds the cumulative record of the total amount of use for each tool so that actual use and life-expectancy figures can be continuously compared.

Using these data files, the computer control component performs a variety of tasks, all of which fall into four categories (Figure 7-6):

1. Program storage and distribution
2. System control
3. System monitoring
4. Reporting

Program Storage and Distribution

A program is required for each set of operations that takes place at each workstation on each part. The computer control component stores these programs and distributes them to the various machines in the system as needed.

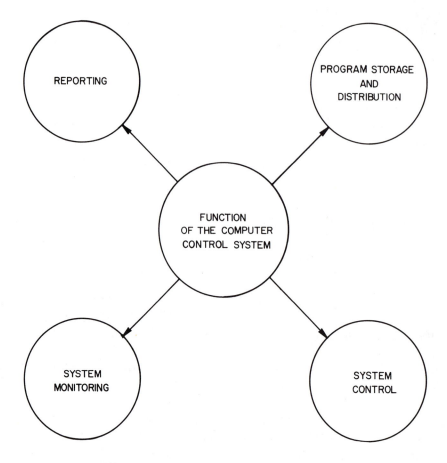

FIGURE 7-6 Function of the computer control system.

System Control

The computer control component provides production, traffic, shuttle, and tool control. Production control involves making decisions about the rate of input of the various workpieces fed into the system and the mix of these workpieces. Traffic and shuttle control involve control of the materials handling component of the system. Tool control involves control of the location of all tools in the system.

System Monitoring

The computer control component monitors all aspects of system operation. It monitors tool life by comparing the expected life of each tool with the cumulative total amount of actual use. It also monitors system performance.

Reporting

The computer control component of the system can produce a variety of reports as specified by manufacturing personnel. Some of the more frequently required reports include status, production, and utilization reports.

Machine Tool Component

The machine tool component of a CIM system consists of the various CNC machines required by the application in question. Some systems incorporate standard CNC machines, some use special machines, and some use both. The type and mix of CNC machines is determined by the needs of the individual manufacturing setting.

Materials Handling Component

The materials handling component of a CIM system moves workpieces from workstation to workstation and orients them at each workstation. Often the materials handling component of a CIM system will actually be two materials handling systems: one for moving workpieces from station to station and one for orienting workpieces at each station.

The system which moves parts is considered the primary system. That which orients parts is considered the secondary system. A secondary system is sometimes referred to as the shuttle system. The primary system is usually a roller or conveyor system. There is a shuttle system at each machine for taking palletized parts from the primary system, placing them on the machine, orienting them properly, and removing them from the machine after processing.

CIM SYSTEM APPLICATIONS

CIM systems are especially designed for the specific manufacturing settings where they will be used. Consequently, there are as many different CIM systems as there are applications. Examples of some actual applications of CIM systems follow.

Figure 7-7 is a CIM system developed by Cincinnati Milacron for Vought Aerospace in Dallas, Texas. The system consists of ten principal stations as shown. The system was installed in 1984 at a cost of $10 million. It is used to produce selected sections of the fuselage for the B-1 bomber manufactured by Vought Aerospace for the U.S. Air Force.

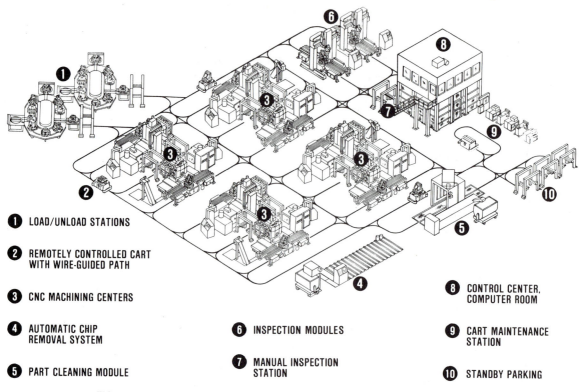

1 **LOAD/UNLOAD STATIONS**

2 **REMOTELY CONTROLLED CART WITH WIRE-GUIDED PATH**

3 **CNC MACHINING CENTERS**

4 **AUTOMATIC CHIP REMOVAL SYSTEM**

5 **PART CLEANING MODULE**

6 **INSPECTION MODULES**

7 **MANUAL INSPECTION STATION**

8 **CONTROL CENTER, COMPUTER ROOM**

9 **CART MAINTENANCE STATION**

10 **STANDBY PARKING**

FIGURE 7-7 CIM system used by Vought Aerospace. Courtesy of Cincinnati Milacron.

In their project projections, Vought personnel estimated the B-1 job would require 200,000 man-hours if undertaken manually. If undertaken using the CIM system shown in Figure 7-7, the manpower projections could be reduced by 130,000, for a savings of $25 million.

Figure 7-8 is a CIM system developed by Cincinnati Milacron for the FMC Corporation in Aiken, South Carolina. The system consists of eight principal stations as shown. The system was installed at a cost of $8 million. It is used to produce selected parts for the Bradley Fighting Vehicle and Multiple Launch Rocket system manufactured by FMC for the U.S. Army.

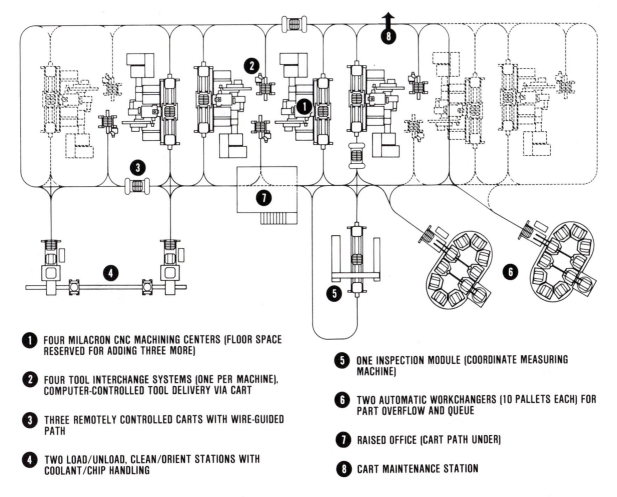

1 FOUR MILACRON CNC MACHINING CENTERS (FLOOR SPACE RESERVED FOR ADDING THREE MORE)

2 FOUR TOOL INTERCHANGE SYSTEMS (ONE PER MACHINE), COMPUTER-CONTROLLED TOOL DELIVERY VIA CART

3 THREE REMOTELY CONTROLLED CARTS WITH WIRE-GUIDED PATH

4 TWO LOAD/UNLOAD, CLEAN/ORIENT STATIONS WITH COOLANT/CHIP HANDLING

5 ONE INSPECTION MODULE (COORDINATE MEASURING MACHINE)

6 TWO AUTOMATIC WORKCHANGERS (10 PALLETS EACH) FOR PART OVERFLOW AND QUEUE

7 RAISED OFFICE (CART PATH UNDER)

8 CART MAINTENANCE STATION

FIGURE 7-8 CIM system used by FMC Corporation. Courtesy of Cincinnati Milacron.

Figure 7-9 is a CIM system produced by Cincinnati Milacron for its own plastics machinery division in Mt. Orab, Ohio. The system consists of thirteen principal stations as shown. It is used to produce seventy-one different machine parts made of cast iron or steel. This particular CIM system reduces part queue times from as high as twenty days to as low as two or three hours.

1 Four Milacron T-30 CNC Machining Centers

2 Four tool interchange stations, one per machine, for tool storage chain delivery via computer-controlled cart

3 Three computer-controlled carts, with wire-guided path

4 Cart maintenance station

5 Parts wash station, automatic handling

6 Automatic Workchanger (10 pallets) for online pallet queue

7 One inspection module — horizontal type coordinate measuring machine

8 Three queue stations for tool delivery chains

9 Tool delivery chain load/unload station

10 Four part load/unload stations

11 Pallet/fixture build station

12 Control center, computer room (elevated)

13 Centralized chip/coolant collection/recovery system (----- flume path)

Cart turnaround station (up to 360° around its own axis)

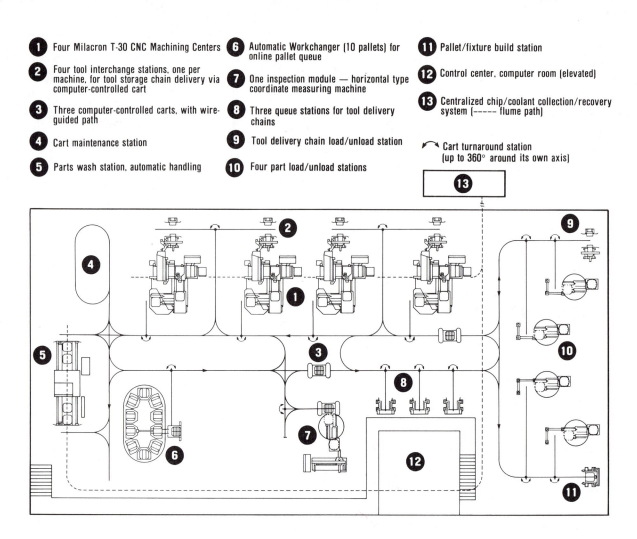

FIGURE 7-9 CIM system used by Cincinnati Milacron. Courtesy of Cincinnati Milacron.

CIM AND PEOPLE

CIM represents a major step toward the totally automated factory. Because of this, human involvement in CIM systems is sometimes overlooked. This should not be; the human component is still an important part of a CIM system.

Like any electromechanical system, a CIM system must be controlled, serviced, and occasionally repaired. These tasks are accomplished by people. People who work with CIM systems fall into four broad occupational categories (Figure 7-10):

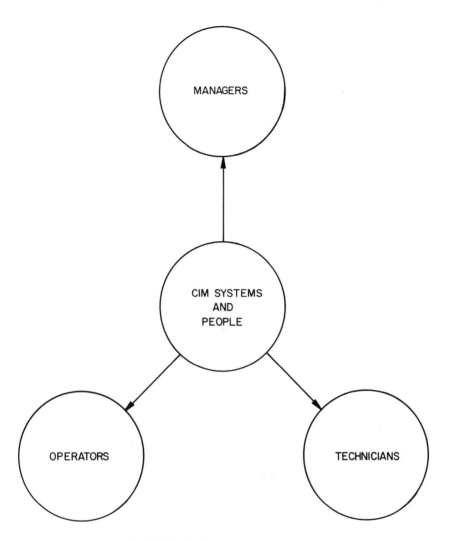

FIGURE 7-10 CIM systems and people.

1. Managers
2. Technicians
3. Operators
4. Programmers

CIM Managers

A CIM system requires a person to see to the overall management of the system. The manager supervises other people involved in the operation of the system and all system operations. The CIM manager is also the decision-maker when exceptions to normal operations are requested. The manager is responsible for production planning and, on occasion, some of the other tasks listed below under the headings "technicians" and "operators."

Technicians

Technicians are responsible for maintaining, servicing, diagnosing, and repairing the various components of the CIM system. These components might include CNC machines, materials handling systems and subsystems, robots, and controllers. This means that mechanical pneumatic, hydraulic, electrical, and electronics technicians are needed with CIM systems.

Operators

CIM system operators must be jack-of-all-trades operators able to perform a variety of tasks. These tasks include preparing tools for use, setting up pallets and fixtures, loading raw stock onto the system, adjusting tools, replacing broken or worn tools, and unloading finished parts from the system.

Programmers

CIM systems depend on computers and controllers in all aspects of their operation. Consequently, programmers are needed. These positions include part programmers as well as regular computer programmers.

PROJECTED TRENDS IN CIM/FMS TECHNOLOGY

The current status of CIM and FMS gives a glimpse of how manufacturing will be accomplished in the future. However, these things are really still in the infancy stage. There are a number of trends already underway which, as they occur, will move CIM/FMS ever closer to becoming the norm in manufacturing. Below is a checklist of projected trends in CIM/FMS technology.

CHECKLIST OF
PROJECTED TRENDS IN CIM/FMS TECHNOLOGY

1. Machine adjustable, tool-changer compatible tools.
2. Automatic tool loading/unloading into tool-changer magazines from the tool crib.

3. Automatic tool crib operation and connection to FMS.
4. Automatic tool exchange between machines.
5. Compact, large capacity, high-speed, random-access, tool-changer magazines.
6. Tool wear and breakage sensors and algorithms.
7. On-line inspection.
8. On-machine inspection.
9. Adaptive error determination and compensation systems.
10. Tighter coupling of CAD and CAM.
11. More comprehensive and data-accessible manufacturing monitoring systems.
12. Comprehensive production planning and management systems to handle inventory, ordering, scheduling, etc.
13. Trouble-free shut-down of system.
14. Automatic generation of system initialization and start-up procedures.
15. Automatic part fixturing and defixturing.
16. Automatic pallet and fixture storage and retrieval.
17. Automatic identification and tracking of tools, pallets, fixtures, parts, carts during all phases of FMS operation.
18. Automated integration of more classes of manufacturing operations.
19. Improvements in automatic chip flushing and clearing, part cleaning, chip collection, and reclaiming and coolant reconditioning.
20. Improvements in automatic temperature control of parts, pallets, fixtures, machines, and coolants.
21. Less roughing required and higher-precision machining.
22. Better equipment maintainability.
23. Increasingly larger FMS's with better integration into the factor planning and operation systems (MRP, etc.).

Courtesy of Wallace Pelton, Texas State Technical Institute.

SUMMARY

Computer integrated manufacturing is what results when individual CAD/CAM concepts are integrated. The term flexible manufacturing system (FMS) is sometimes used synonomously with CIM, but in reality flexible manufacturing systems are one type of CIM system.

CIM is, in reality, a modern version of the manufacturing integration that existed in the days of the early craftsman who accomplished all phases of production. Such integration was lost when, as part of the industrial revolution, specialization became the norm in manufacturing.

There are as many different types of CIM systems as there are different manufacturing settings, because each system is especially designed for a given setting.

CIM came about in response to the long-standing need for a manufacturing capability between stand-alone machines on the one extreme and transfer lines on the other. CIM systems fill this need for a medium-volume, moderate-flexibility capability in manufacturing.

CIM offers a number of advantages, including (1) ability to produce families of parts, (2) accommodation of the introduction of random parts, (3) requirement of less lead time, (4) allowance for a closer relationship between parts to be produced and in-stock inventory, (5) better machine utilization, and (6) requirement of less labor.

There is resistance to CIM at both the management and worker levels. This is partially because the need for CIM has been seen at the shop floor level. Consequently, interest has been from the bottom up rather than from the top down.

All of the many different configurations of CIM systems fall into one of three groups: (1) special systems, (2) flexible manufacturing systems, and (3) manufacturing cells. Of these, those which fall into the first group are the least flexible but have the highest production rate. Those in the second group have more flexibility but a lower production rate. Those in the third group are very flexible but have a correspondingly limited production rate.

All CIM systems have three components: (1) computer control component, (2) machine tool component, and (3) a materials handling component. Although people are not normally associated with CIM, CIM systems do require people in order to operate. These people fall into one or more of four catagories: managers, technicians, operators, and programmers.

Chapter Seven REVIEW

1. Define the term computer integrated manufacturing.
2. How does CIM differ from FMS?
3. Explain the three ways in which a computer is used in a CIM system.
4. Explain the rationale for CIM systems.
5. Explain why there is resistance to CIM.
6. What are the three categories of CIM systems?
7. What are the components of a CIM system?
8. List five different types of data files used in CIM systems.
9. What types of positions make up the "people component" of a CIM system?

The problem of job displacement is representative of the types of problems CAD/CAM managers face on a daily basis. Knowing how to deal with them properly can be the difference between the success or failure of a CAD/CAM operation.

Thus far, each chapter has dealt with the technical aspects of a given CAD/CAM concept. Students of CAD/CAM need to be well versed with regard to such technical information. However, they also need to be familiar with CAD/CAM from a manager's perspective. Many students completing post-secondary programs of study relating to CAD/CAM will become CAD/CAM managers during their careers.

In addition to the technical knowledge about CAD/CAM, managers need to know how to decide whether a conversion to CAD/CAM is appropriate; how to select a vendor or vendors; how to evaluate the hardware, software, and services of a vendor; and what personnel problems might arise and how to deal with them.

Major Topics Covered

- CAD/CAM: To Convert or Not To Convert
- Evaluating Vendors and Selecting Systems
- Personnel Concerns

Chapter Eight

CAD/CAM Management

Technology is too important a resource to be ignored in our strategic business planning!" Bob Adams, President of Aerospace Products, stated at his Board Meeting.

Aerospace Products, a pseudonym for a medium-sized aerospace subcontractor, had an international reputation for providing quality engineered products. However, like most industries, the company was experiencing significant cost pressures from competitors. At the same time, commercial and military customers were demanding shorter lead times, better product reliability, and lower unit costs. While the company was investing millions of dollars annually in new technologies, these investments lacked a clear, integrated direction. As one executive termed it, "We're creating too many islands of technology." Also, lacking a clear picture of the financial returns from such investments, management was reluctant to invest in technology "for technology's sake."

Upon further discussion at the Board of Directors' meeting, the need for a well-designed business strategy for implementing Computer-Integrated Manufacturing (CIM) technologies became readily apparent. There was no doubt in anyone's mind that the harnessing of the right material, process, and information technologies would provide Aerospace Products with a significant competitive edge. However, the central issue was one of determining how substantial new technology investments could be put to the most effective use for meeting company goals and long-term objectives. Some of the strategic questions regarding the implementation of new technologies that management wanted to address were:

- What business functions would benefit most from new technologies?
- How can we ensure an adequate return on our investments in capital-intensive technologies?
- How should we integrate different technologies into a true CIM environment?
- How can we implement short-term improvements for immediate results while we wait for longer-term technology solutions?

Courtesy of SME Book: A Program Guide for CIM Implementation, First Edition, Third Printing.

- How should we translate a high-level CIM business strategy into a tactical plan for specific actions?
- How can we effectively manage the inherent risks associated with the implementation of new technologies?

To develop a CIM business strategy, Aerospace Products engaged the services of an outside management consultant. The consultant provided the necessary initial and day-to-day project management direction, and also provided specialized techniques, education, and training to Aerospace Products personnel in the development of this strategy. Company personnel, in turn, were heavily involved in all facets of the project, so that they could effectively implement the tactical plans developed and continue to maintain the CIM business strategy as a living document.

In the remainder of this paper, a synopsis of the approach used for developing the CIM business strategy is presented.

The company's established Strategic Business Plan was used to provide a long-term outlook for future products, markets, and business environments. Based on this outlook, a high-level Capital Investment Strategy was developed to handle the realities of evaluating new technology investments. A well-defined project management approach set the stage for controlling the different activities required to develop the CIM business strategy. Using the Capital Investment Strategy as a guideline, a Needs Analysis was conducted to determine what business functions could most effectively profit from new technologies. A key element of the Needs Analysis was the development of a Cost Model that defined the "real world" costs associated with different business functions by cost element. This Cost Model was used in all phases of the strategy.

Once the Needs Analysis was completed and specific company functions defined as the best candidates for technology modernization, the conceptual design activities were initiated. During this stage, alternative technologies were evaluated and selected to overcome major deterrents to productivity in the current business environment. These technologies

(continued)

were evaluated for cost/benefit purposes using the Cost Model. Finally, the individual technology projects were converted into a Master Plan for implementation. This Master Plan would take the company from its current business environment to its required CIM environment over a period of several years. Finally, mechanisms were established to monitor the implementation of the Master Plan on a project-by-project basis. This would allow project management to fine tune the CIM business strategy on a continuing basis.

This overall CIM business strategy framework was established at Aerospace Products over a period of eight months. Its implementation can best be described in four discrete stages:

- Stage 1: Project initiation
- Stage 2: Needs analysis
- Stage 3: Conceptual design
- Stage 4: Master plan

The following is a review of the key activities that took place in each stage to make the CIM business strategy a reality.

STAGE 1: PROJECT INITIATION

During this stage, the overall project organization and methodologies were established. Personnel were assigned to the project and trained in specialized techniques for developing the CIM business strategy. The Corporate Investment Strategy was examined and refined to meet the company's changing needs.

Project Management Plan

One of the first activities undertaken was the development of a Project Management Plan. This plan clearly defined the key tasks to be undertaken for the development of the business strategy. Expected results and deliverables, estimated man hours by specific personnel assigned, and planned completion dates were defined for each key task. The plan formed the basis for allocating resources and managing the project to ensure a timely and quality business strategy.

Project Organization

A high-level steering committee was appointed to provide top management direction and commit resources to the CIM business strategy development project. It consisted of the President and the Vice President in charge of each major organizational function.

A Project Team consisting of selected professionals from industrial engineering, management information systems, design engineering, manufacturing engineering, advanced technologies, finance, materials management, manufacturing, purchasing, and quality assurance was assigned to the project based on the needs expressed in the Project Management Plan.

A project leader was selected to head the project team. Appropriate periodic reporting mechanisms were established to provide sound communications among top management, project team personnel, and key middle managers in the organization.

Capital Investment Strategy

An evaluation of the company's investment strategy showed that the direct labor-based cost accounting system used to justify capital expenditures was inadequate for the financial analysis and justification of new technologies. While direct labor costs were very well-defined by specific manufacturing function and product, they represented less than 10% of total manufacturing costs. On the other hand, key costs, such as those

(continued)

associated with carrying inventories, scrap, tooling, computer hardware and software development, were typically buried in large overhead pools. As a result, no common yardstick existed for evaluating the cost/benefit impact of different technologies on significant cost elements within a business function.

With the active involvement of the Steering Committee, a new Capital Investment Strategy was developed to meet the future needs of the organization. The five-year strategic business plan was used as a basis for defining the future products and business environment in which Aerospace Products would operate. The company's critical success factors were then defined. Appropriate performance measures were developed for each critical success factor. These performance measures were classified as financial and nonfinancial. The financial performance measures were organized by significant cost element into a cost model (discussed later). A weight system was assigned to financial and nonfinancial measures for developing a ranking methodology. This methodology formed the basis for comparing the cost/benefit profiles of different technologies, based on top management's critical success factors.

Also, through the Capital Investment Strategy, corporate guidelines were established for defining individual technology projects on a consistent basis; identifying cost/benefits for each project; obtaining technical, financial and management approvals; and determining the minimum returns on investment expected by management for authorizing capital investment.

By participating actively in the development of the Capital Investment Strategy, the Steering Committee established a meaningful "top-down" commitment to the entire project, as top management got an excellent perspective of the changes in the processes needed for implementing modern technologies.

Education and Training

As a final step of the project initiation process, a number of seminars were conducted for Aerospace Products personnel. Project plans, resources required, and results anticipated were reviewed with all top management, project team, and key middle management personnel throughout the company. Technical seminars on specific techniques were held for project team members to provide them with the necessary tools for conducting the project. These included methodologies for conducting factory analysis, development of a cost model for financial analysis, evaluating different business functions for improvement potential, developing conceptual designs for individual technologies, conducting return-on-investment analysis utilizing the Capital Investment Strategy guidelines, and the overall project management methodologies for ensuring high-quality results in a timely manner.

Having completed project initiation, the Aerospace Products project team was now ready to start on the Needs Analysis.

STAGE 2: NEEDS ANALYSIS

The Needs Analysis stage was aimed at analyzing the company's business functions to determine where technology could be applied most effectively, based on management's objectives as established in the Capital Investment Strategy.

Factory Model

Using a structured modeling approach, the Aerospace Products organization was analyzed by major business function such as marketing, product design, production, manufacturing engineering, materials control, quality assurance, and other administrative support functions. Using smaller

(continued)

study groups within the overall project team, each of these major business functions were decomposed into subfunctions at greater and greater levels of detail. Through this analysis, the organization was subdivided into a total of 416 functions. These functions ranged from shop work-centers, such as a group of vertical broaching machines, to indirect functions, such as master production scheduling.

The result of this analysis was a factory model that clearly showed the decomposition of major functions into well-defined subfunctions. The model also illustrated the interrelationships between the 416 functions.

Cost Model

Using special computerized models, business costs for the current year were analyzed in a top-down manner through the factory model. Fourteen cost elements were defined to be truly significant across the organization based on the Capital Investment Strategy. These cost elements were different from those used in the traditional cost accounting system. For example, they include inventory carrying costs, scrap, rework, energy, tooling, and information systems.

Controllable business costs for each major business function were systematically allocated in a top-down manner to subfunctions in the factory model based on available financial and statistical data. Throughout the process, care was taken to ensure that each function's cost was reasonably allocated and that, when combined, the total cost of all functions was equal to the overall controllable business costs for the organization.

The result of this analysis was a cost model that reasonably depicted the current costs associated with each of the 416 functions constituting the organization. Further, it provided a breakdown of costs across the 14 cost elements for each of the 416 functions.

Improvement Potential Analysis

Using a specially designed structured approach, project team members then assessed the improvement potential for each function. This was based on an evaluation of current materials, methods, processes, and information systems. Current performance was analyzed based on quantitative and subjective data. Available technologies were considered. Also, a variety of organizational, training, policy, procedural, methods, and system enhancement issues surfaced.

As a result of this analysis, a list of 120 recommendations were made to management for short-term improvements not requiring major capital investments. Over half these recommendations were implemented within the next six months. resulting in significant improvements in work flow, product reliability, and employee morale. In addition, a quantitative improvement potential assessment was made for those functions that could be significantly aided through longer-term material, process, or information technologies.

Prioritizing Improvement Opportunities

By applying the results of the improvement potential analysis to the costs associated with each function in the cost model, functions with the best opportunities for utilization of new technologies were identified.

High-cost improvement functions were then evaluated for integration into technology areas (Tech Areas) based on work cell/center combinations of functions as well as the linking of direct and indirect functions. Also, nonfinancial criteria were evaluated for each major Tech Area. Tech Areas were then ranked by priority, using the ranking methodology developed as part of the Capital Investment Strategy.

As a result of this process, four Tech Areas were selected by the Steering Committee for further development into specific technology projects for conceptual design.

(continued)

STAGE 3: CONCEPTUAL DESIGN

During the Conceptual Design stage, the four Tech Areas selected were analyzed in depth for technology modernization. Based on the research conducted, conceptual designs for the most feasible integrated technology projects were developed. The cost/benefit analysis, for each individual project and for all the individual projects as a whole, was evaluated taking into account technology, and human and financial risks associated with each individual project. The total package of suitable technology projects then was presented to the Steering Committee for review and approval.

Technology Projects Definition

Each Tech Area was carefully analyzed to identify the significant productivity drivers associated with each cost element of the functions included in the Tech Area. For example, the impact of scrap in one fabrication Tech Area was found to be approximately $500,000. Analysis of the factors causing the scrap showed that 50% of the scrap was traceable to defective tooling design, while 30% was traceable to inadequate process control.

Potential technology alternatives were researched by the project team with the assistance of appropriate technical experts. The key objective of the project team was to identify those alternatives that could most significantly remove the deterrents to maximizing cost reduction, product reliability, and schedule performance. The cost model was used as a basis for simulating the impact on Tech Area costs associated with each technology alternative.

By eliminating marginal solutions and combining different technology alternatives, a selected few technology projects were identified for each Tech Area. Conceptual designs were prepared for each project.

Implementation Analysis

The detailed design and implementation activities associated with each technology project were defined. A preliminary cost/benefit analysis was prepared using the cost model. Care was taken to ensure that benefits were not duplicated for different technologies applied to the same function. Assumptions regarding technical considerations and financial analysis were carefully recorded, to support the implementation analysis.

Risk Assessment

At this point, each technology project was presented by the project designer to key users who would be responsible for obtaining the results anticipated. The purpose of these discussions was to obtain constructive criticism regarding technical, human, and financial assumptions surrounding the project. Sensitive assumptions were carefully tested by obtaining the viewpoints of those who would have to "live" with the solutions.

As a result of this step, the conceptual designs and their cost/benefit analysis were modified based on new knowledge gained through the meetings. Also a formal "risk management" program was prepared for each project to define how the design and implementation of each technology project would be managed to maximize the real benefits to the company. Besides the significant technical and cost-validation advantages obtained, this step resulted in a very positive acceptance of responsibility by key middle managers to "make the projects happen."

Finalizing Conceptual Designs

Project team members then finalized their conceptual design documents for each project, ensuring that besides the technical designs

(continued)

they also had all relevant data on strategies and costs for detailed design and implementation of the individual technologies along with anticipated benefits. The tracking mechanisms for monitoring schedule, cost and benefit performance also were defined for each project.

Because of the variety of material, and process and information technology projects involved, care was taken to ensure that the design documents prepared were consistent with the guidelines established in the Capital Investment Strategy.

Project Economics

A financial analysis was then conducted to identify the return on investment for each project and all the projects as a whole. Based on the Capital Investment Strategy, this included assessments of before- and after-tax implications, investment tax credit considerations, and savings/ investment ratios. The return on investment was compared to the company-established hurdle rates. The overall package was then sent to management for approval.

As a result of the conceptual design process, twelve technology projects were identified for implementation. Some representative ones were

- a Flexible Manufacturing System (FMS) for a fabrication area,
- a robotic welding line,
- an Automated Guidance Vehicle System (AGVS),
- an integrated engineering database for CAD/CAM/CAE,
- a process monitoring system, and
- a modified cost management system.

STAGE 4: THE MASTER PLAN

The fourth and final stage in the development of the CIM Business Strategy was the organization of the approved conceptual design projects into a suitable master plan for implementation at Aerospace Products.

Implementation Plan

The individual technology projects were analyzed for logical interdependencies from a CIM environment perspective. Because Aerospace Products desired to move quickly to a flexible manufacturing environment, it was vital that the master plan accommodate the progressive union of the different technology projects into an integral long-term solution. Certain projects, such as the integrated engineering database, had therefore to precede other dependent technology projects, so that a suitable architecture was in place for the longer-term CIM environment. Given the individual cost/benefit assessments by project, as well as the combined picture for all projects, top management was in a better position to make meaningful decisions on the master planning process.

Also, a resource profile was constructed to identify the types and quantities of skills needed for implementing the individual technology projects as well as the capital investments associated with each project. Based on management's willingness to commit annual capital funds to the CIM business strategy, the master plan was finalized. It detailed specific steps, personnel responsbilities, and key time periods for accomplishing results against the overall master plan for technology modernization.

CIM Business Strategy Maintenance

Organizational mechanisms for monitoring progress against the Master Plan were established. Management recognized the need for tracking costs and benefits at a reasonable level of detail by specific project. For the CIM business strategy to be a "living document," it was essential that actual variances from planned expectations be fed back to the steering committee on a periodic basis. This information, along with an analysis of reasons for variance, could then be used to take appropriate corrective measures by either fixing problem areas or changing the original assumptions for each project.

(continued)

RESULTS OBTAINED

What were the key results obtained by Aerospace Products through this project?

Five of the most significant results were as follows:

- Top management was able to answer the strategic questions that it had regarding new technologies. The result: a commitment to invest over $28 million in CIM Implementation over a five-year period based on a reasonable projection of return on investment and other competitive advantages.

- There was a significant consensus of opinion at all management levels that the right technology projects had been selected. Further, key managers and professionals had a strong sense of commitment to a successful implementation of individual projects because of their involvement with the selection and design process.

- Different professionals with engineering, manufacturing, information systems, and financial backgrounds developed a new-found respect for one another's abilities by working to a common goal through the project team approach. This boosted morale and was expected to have a favorable impact on the productivity of these professionals.

- The cost model provided a useful mechanism for analyzing the impact of all capital investment projects, not just those associated with CIM technologies. Management was particularly impressed with its practical ability to measure indirect costs and benefits, avoid overstatement of benefits, and analyze the impact of different cost drivers.

- Finally, and most importantly, management had successfully incorporated technology integration into its overall strategic business planning process. This gave Aerospace Products a significant competitive edge for increasing corporate return on investment and more effectively serving its customers, while developing a flexible manufacturing environment.

The problem of job displacement is representative of the types of problems CAD/CAM managers face on a daily basis. Knowing how to deal with them properly can be the difference between the success or failure of a CAD/CAM operation.

Thus far, each chapter has dealt with the technical aspects of a given CAD/CAM concept. Students of CAD/CAM need to be well versed with regard to such technical information. However, they also need to be familiar with CAD/CAM from a manager's perspective. Many students completing post-secondary programs of study relating to CAD/CAM will become CAD/CAM managers during their careers.

In addition to the technical knowledge about CAD/CAM, managers need to know how to decide whether a conversion to CAD/CAM is appropriate; how to select a vendor or vendors; how to evaluate the hardware, software, and services of a vendor; and what personnel problems might arise and how to deal with them.

CAD/CAM: TO CONVERT OR NOT TO CONVERT

The most fundamental management concern with regard to CAD/CAM is whether or not to make the conversion. This can be a difficult decision because there are so many unknowns and uncontrollable variables involved. However, if one takes a step-by-step, systematic approach, many of the unknowns can be eliminated and a well-informed decision can be made. In making such important decisions, it is critical to have the best information possible. The manager can obtain such information by

1. identifying companies similar in size, products, and configuration that have already converted to CAD/CAM and discussing the process with their personnel;
2. analyzing their company's financial forecast; and
3. becoming familiar with factors which indicate the need for a conversion and those which indicate otherwise.

Contacting Similar Companies

Managers are generally aware of other companies which are similar to theirs in size, products, and overall configuration. By identifying their counterparts in such companies, CAD/CAM managers can obtain valuable first-hand information concerning the advisability of a conversion. The types of questions that should be asked include:

1. What vendor or vendors did you deal with? Were they satisfactory? What problems did you have with the vendor(s)?
2. Did you hire new employees or retrain existing workers?
3. What impact did the conversion have on employee morale?
4. Did productivity decline after the conversion? If so, for how long before it began to climb again?
5. What actual productivity gains have occurred?
6. What were your initial costs? What are your ongoing costs? What hidden costs have been detected?
7. What strategies is your company using to make CAD/CAM systems more cost effective?
8. What benefits have your realized? Planned? Unplanned?
9. If you had the process to undertake again, would you do it? If so, what would you do differently?

Analyzing the Financial Forecast

Financial forecasting is a normal part of business in most companies. Before making a decision as to whether a CAD/CAM conversion is advisable, the financial forecast for the subject company should be analyzed carefully. Using this document as the starting point, managers can calculate the expected costs for continuing in the manual format and for converting to the CAD/CAM approach. These calculations then can be compared.

Normally, a company that is forecasting an increased workload will be a candidate for a conversion to CAD/CAM. Those that project level growth or a decline are not normally good candidates. However, there are exceptions to this rule of thumb. One is the case of companies which will increase their workload potential by converting to CAD/CAM. An example of this would be a company which is losing contracts because bid specifications require that all documentation be done on a CAD system or that tolerances be maintained that are not realistic on manually controlled machines.

Another case in which a CAD/CAM conversion may be appropriate even when facing level or declining growth is when such declines are to be accompanied by corresponding reductions in the work force. When this is the case, managers may elect to marry the expected reduction in force with a CAD/CAM conversion.

Forecasted growth is an indictor that a CAD/CAM conversion may be warranted, but it is not a definitive enough criterion, by itself, upon which to base a decision. Rather, it indicates that there is potential and the matter should be studied further. The next step is to compare the cost estimates for the manual and CAD/CAM approaches.

Estimating the Cost of the Manual Approach

Regardless of the type of operation in question (manufacturing, design, etc.) growth in the workload will result in some predictable needs that will have corresponding costs. These costs can be estimated with a high degree of accuracy. The needs fall into four broad categories:

1. Personnel needs
2. Facility needs
3. Furniture and equipment needs
4. Expendable supply needs

In projecting new personnel costs, managers should consider the following factors:

1. What types of positions will be needed (i.e., designers, drafters, machinists, assemblers, etc.)
2. How many man-hours or man-years are projected for each type of position?
3. What are the projected wage rates for each type of position.
4. What are the benefit costs associated with each type of position?

Figure 8-1 is a forecasting worksheet for estimating the personnel costs which will result from a projected growth over a specified period. How this worksheet might be used can be seen in the following example.

SAMPLE CALCULATION

The forecast for a given company projects a significant increase in its workload. The increased workload will mean that additional personnel will be required in five types of positions: designers, drafters, machinists, assemblers, and office clerks. Costs estimates are to be accomplished for a five-year period. It is estimated that designers will be needed for 30,000 man-hours over the five-year period. The average wage rate will be $26.75 per hour. Another 20% must be added to cover the cost of benefits. The computation would be accomplished as follows:

Step 1: For convenience, convert the man-hour figure to man-years:

30,000 man-hours

Step 2: Since man-years are to be used, the equivalent wage rate must be determined:

2,000 hours per year X $26.75 per hour = $53,500.

Time Period _____

Type of Position	*Man-Hours Projected	*Wage Rates	Benefit Costs	Subtotals
Sub-Totals				

COMPUTATION PROCEDURE
(Type of Position) Man-Hours X Wage Rate + Benefit Costs = Sub-Total

*Man-hours can be converted to man-years for convenience with a corresponding conversion of wage rates. Unless individual company policies specify otherwise, use 2,000 man-hours per man-year. Multiply the hourly wage rate times 2,000 to determine the man-year equivalent wage rate.

FIGURE 8-1 Forecasting worksheet for estimating new personnel costs.

Step 3: The man-years are multiplied times the man-year wage equivalent:

$53,500 cost per man-year X 15 man-years = $802,500.

Step 4: The benefit costs may now be computed. Unless company practices specify otherwise, increase the amount from Step 3 by 20%:

$802,500.00 Cost/15 man-years
X   1.20 Benefit factor
$963,000.00

Based on these calculations, the company can expect to incur personnel costs of $963,000 for the additional design positions that will be needed over the five-year period. Using the same approach, the costs of the other four types of positions can also be estimated. The next step is to estimate the facility-related costs.

In determining the facility needs which will result from projected growth in the workload, three questions should be asked:

1. Will new construction be required?
2. Will renovations to existing facilities be required?
3. Will additional facilities be rented or leased?

If new construction will be required to accommodate the personnel who will be added, the costs associated with it must be estimated. This involves identifying the types of space that will be needed (i.e., office, storage, design and drafting, manufacturing, etc.) as well as the amount of gross square feet required for each type. With this done, the average per square foot figure can be applied to determine the estimated cost of new construction. Figure 8-2 is a forecasting worksheet for estimating the cost of new construction.

Type of Space	Net Sq.Ft. Required	*Gross Sq.Ft. Required	Avg. Cost Per Sq.Ft.	Subtotal
_____	_____	_____	_____	_____
_____	_____	_____	_____	_____
_____	_____	_____	_____	_____
_____	_____	_____	_____	_____
_____	_____	_____	_____	_____
_____	_____	_____	_____	_____
_____	_____	_____	_____	_____

COMPUTATION PROCEDURE

1. Net Square Feet Required X *1.30 = Gross Square Feet Required

2. Gross Square Feet Required X Average Cost Per Square Foot = Subtotal

*Gross square footage is the net square footage plus allowances for interior heating and cooling equipment, electrical equipment, hallways, and other support areas. It can be computed by multiplying the net square footage times a constant factor of 1.3 unless local policies and practices specify otherwise.

FIGURE 8-2 Forecasting worksheet for estimating new construction costs.

How this worksheet might be used can be seen in the following example:

SAMPLE CALCULATION

A forecasted increase in workload will require new construction at a given company's main site. It has been determined that the following types of spaces will be needed in the specified amounts:

Design/drafting space .... 1,800 square feet
Office space .... 300 square feet
Storage space .... 200 square feet

These amounts represent net square footage, which does not include allowances for support areas such as inside heating and cooling units, plumbing, hallways, electrical equipment, and so on. A local architect has advised that $88 per square foot can be used for the average cost figure. The computation would be accomplished as follows:

Step 1: Since the $88 per square foot can be applied to all three types of areas, the total net square footage should be calculated:

1,800 square feet
300 square feet
<u>200</u> square feet
2,300 square feet Total Net Square Footage

Step 2: The total net square footage figure should be increased by 30% to determine the gross square required:

2,300 net square feet X 1.30 = 2,990 gross square feet

Step 3: The gross square footage figure is multiplied times the average cost per square foot:

2,990 gross square feet X $88 per square foot = $263,120

If renovations to existing facilites are needed, managers must depend on an architect or contractor to provide estimates. Unlike new construction, renovations cannot be assigned an average cost per square foot figure because the types of activities required can vary so drastically. For example, the renovations might range in complexity from something as minor as repainting to something as involved as gutting a building and reconstructing the interior.

If facilities are to be leased, projecting the cost is a matter of multiplying the yearly rate times the number of years the facilities will be needed. Managers should be sure to allow for the annual increases which can be expected in lease contracts.

In determining new equipment costs for the manual option, managers have the advantage of knowing what types of equipment are needed to support existing positions. With this knowledge, it is a simple process to project the types and amounts of equipment that will be needed to support new personnel.

However, one must be sure that allowances are made for expected future increases or decreases in the costs of such equipment. It is not a sound practice to use today's prices in projecting the cost of equipment that will be purchased at a point in the future.

Equipment vendors can provide accurate data about the future costs of their products. By requesting this type of information from several competing vendors, managers can obtain accurate data on which to base their projections. The same approach can be used in projecting the costs of expendable supplies. The types and amounts needed to support existing personnel can be used in determining the needs of future personnel.

At this point, the manager can add the estimated costs of personnel, facilities, furniture and equipment, and expendable supplies to determine the overall projected cost of the manual approach. The next step is to develop the cost estimate for the CAD/CAM approach for the same period and based on the same workload projections.

Estimating the Cost of the CAD/CAM Approach

As with the manual approach, the costs of the CAD/CAM option can be estimated with a high degree of accuracy. The company's needs with regard to the CAD/CAM approach fall into eight categories:

1. Personnel costs
2. Initial hardware costs
3. Initial software costs
4. Expansion costs
5. Training costs
6. Maintenance costs
7. Miscellaneous costs
8. Facility-related costs

The costs in each category can be estimated and added together to arrive at an overall estimate of the cost of the CAD/CAM option.

The first task involves estimating the personnel costs. The starting point for accomplishing this task is the forecasting worksheet used to estimate the personnel costs of the manual approach.

If the types of positions on the worksheet include designers and drafters, one should expect to need fewer man-hours or man-years as a result of the productivity gains from CAD. If the types of positions on the worksheet include machinists and machine operators, one should expect to need fewer man-hours or man-years as a result of the productivity gains from CNC or CIM. If the types of positions on the worksheet are low-skill positions, and robots are to be purchased to replace most of them, one should expect to need significantly fewer human man-hours.

Of course, the fundamental task of the manager at this point is to determine the amount of manpower savings which can be expected from projected productivity gains. To do this, managers apply productivity ratios to the personnel projections already estimated for the manual approach.

For example, if, by purchasing a CAD system or CNC machine, one can expect to achieve a 2-to-1 ratio (work that will take two hours to accomplish manually can be accomplished in one hour with CAD or CNC), one can also expect to cut the manpower projections in half. The problem is to determine what productivity ratio to apply.

What is needed at this point are reliable figures for making projections. There are a variety of sources from which to obtain productivity figures. Vendors provide them. The literature is replete with such information. However, the most valid data will probably come from similar companies which have undergone a CAD/CAM conversion and have actual figures on file. By talking with their counterparts in these companies, managers can obtain realistic figures with which to work. Even these should not be viewed as hard and fast, since no two companies are exactly alike and no two companies will achieve exactly the same results.

In talking with veterans of a conversion, managers will learn that productivity gains tend to begin slow and increase over a period of time as the company and its personnel become more accustomed to the CAD/CAM approach. For example, it is not uncommon for a company to show no improvement or even a decline in productivity during the first year of a conversion. By the end of the fifth year, this same company may show gains as high as 4 to 1. Because of this and the many other variables which enter into the equation, managers are well advised to use the "best case-worst case" approach in estimating the savings which will result from a CAD/CAM conversion.

Figure 8-3 is a graph of the best- and worse-case scenarios over a five-year period for a hypothetical design and drafting department. In the worst case, the department will show no improvement in the first

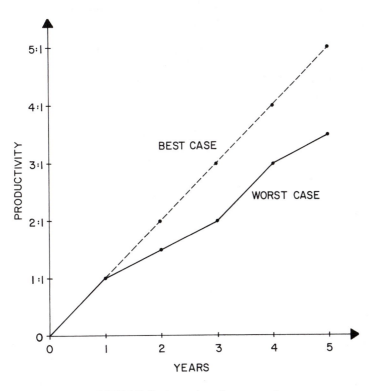

FIGURE 8-3 Best-case/worst-case graph.

year. In year Two, improvement begins to appear. From year Two through year Five the improvements expected for the worst case are:

Year Two	1.6 to 1	Year Four	3.0 to 1
Year Three	2.0 to 1	Year Five	3.6 to 1

Over a period of five years, the CAD/CAM option, in a worst-case scenario, will yield productivity improvements of an average rate of 2.24 to 1. Before applying this figure to the man-year estimates for the manual approach, the figures for the best case scenario should be run.

From Figure 8-3, it can be seen that in year One, as with the worst case, no improvement is projected for the best case. However, by the end of year Two, substantial gains have been made. From year Two through year Five, the improvements expected for the best case are:

Year Two	2 to 1	Year Four	4 to 1
Year Three	3 to 1	Year Five	5 to 1

Over a five-year period, the CAD/CAM option, in the best-case scenario, will yield productivity improvements of an average of 3 to 1. Now the best- and worst-case figures can be applied to determine the savings that can be expected in personnel costs. In the earlier example of the manual approach, it was determined that fifteen man-years of design time would be needed over the five-year period at a cost of $53,500 per man-year.

Figure 8-4 shows how the projected productivity improvements would be applied to these figures for the worst-case scenario. The manual approach will cost the company $802,500 in personnel costs over the five-year period. The CAD/CAM approach, because of the improved productivity, will cost the company $358,450 in personnel cost for the same period. This represents a savings of $444,050 for the worst-case scenario.

Approach	Man-Years	Cost Per Man-Year	Cost
Manual	15.00	$53,500	$802,500
CAD/CAM	*6.70	$53,500	$358,450
		Projected Savings	$444,050

*Man-year equivalent after applying a 2.24 to 1 productivity ratio

FIGURE 8-4 Labor cost comparison worksheet.

Figure 8-5 shows how the projected productivity improvements would be applied for the best-case scenario. In this case, the 3-to-1 ratio reduces the man-year figures to five. This results in personnel costs of $267,500 for the five-year period, or a savings of $535,000.

Approach	Man-Years	Cost Per Man-Year	Cost
Manual	15.00	$53,500	$802,500
CAD/CAM	*5.00	$53,500	$267,500
		Projected Savings	$535,000

*Man-year equivalent after applying a 3 to 1 productivity ratio

FIGURE 8-5 Labor cost comparison worksheet.

When all personnel costs have been projected, managers must estimate the costs that will grow directly out of the purchase of CAD/CAM systems. These costs fall into six categories:

1. Initial hardware costs
2. Initial software costs
3. Expansion costs
4. Maintenance costs
5. Training costs
6. Miscellaneous costs

In arriving at projections of the costs in each of these categories, managers must depend on the estimates of vendors. It is important to ensure that estimates are requested from several vendors, but only from vendors that can actually meet a company's needs.

Initial Hardware Costs: Estimates of the initial hardware costs should include all items that will be needed to completely configure the system, be it a CAD, CNC, robotics, or CIM system. It should include all machines, equipment, furniture, connection cables, and any other devices required to configure the hardware.

Initial Software Costs: There are a number of different types of software available for the various types of CAD/CAM systems. The initial software costs should include all software needed to drive the system. This includes operational, application, and optional packages.

Expansion Costs: During the initial purchase proceedings, a future expansion plan is developed jointly by potential vendors and the buyer. This plan will list the approximate costs of future expansion of the system(s). Such projections should parallel those done for initial hardware purchases. This is to say, they should take into account all hardware items that will be needed to accomplish the planned expansion.

Maintenance Costs: The costs of maintaining the hardware and software over the projected period must be made part of the overall estimate of the CAD/CAM approach. Hardware maintenance agreements take effect after the warranties expire. Software maintenance agreements ensure that buyers receive all updates and improvements to the software they have purchased as they develop.

Training Costs: There will be training costs associated with a conversion to CAD/CAM. Managers and users both will need training that is specifically geared to their needs. With technical personnel or users, there will need to be both beginning and advanced level training. There may or not need to be additional training in the years which follow the initial implementation. This will depend on the degree of change resulting over the years from hardware and software updates. In any case, training costs should be estimated for the entire projection period, even if the projection is zero for a given year or years.

Miscellaneous Costs: Miscellaneous costs include such items as expendable supplies and such services as those provided by consultants. Hidden costs turned up through questioning colleagues who have already gone through a CAD/CAM conversion can also be placed in this category. It is important to estimate the cost of expendable supplies. One might expect that the cost of supplies for the manual operation and for the CAD/CAM option would simply cancel each other out and, therefore, can be left out of comparative projections. However, this is not the case. The types of expendable supplies required with each option are different enough in type and cost that it is wise to include them in the projections.

Figure 8-6 is a worksheet that can be used for estimating the costs directly associated with a CAD/CAM purchase. It is set up for a five-year projection period, but can be shortened or extended as necessary.

INITIAL HARDWARE COSTS

Item	Quantity	Unit Cost	Item Subtotal
_____	_____	_____	_____
_____	_____	_____	_____
_____	_____	_____	_____
_____	_____	_____	_____
		Hardware Total	_____

INITIAL SOFTWARE COSTS

Type	Quantity	Unit Cost	Item Subtotal
_____	_____	_____	_____
_____	_____	_____	_____
_____	_____	_____	_____
_____	_____	_____	_____
		Software Total	_____

MAINTENANCE COSTS

Type	Year 1	Year 2	Year 3	Year 4	Year 5	Subtotal
Hardware	_____	_____	_____	_____	_____	_____
Software	_____	_____	_____	_____	_____	_____

FIGURE 8-6 CAD/CAM purchase cost estimation worksheet.

EXPANSION COSTS

Area	Year 1	Year 2	Year 3	Year 4	Year 5	Subtotal
Hardware	_____	_____	_____	_____	_____	_____

TRAINING COSTS

Type	Year 1	Year 2	Year 3	Year 4	Year 5	Subtotal
Management	_____	_____	_____	_____	_____	_____
User (Begin.)	_____	_____	_____	_____	_____	_____
User (Adv.)	_____	_____	_____	_____	_____	_____

MISCELLANEOUS COSTS

Type	Year	Year 2	Year 3	Year 4	Year 5	
Supplies	_____	_____	_____	_____	_____	_____
Consultants	_____	_____	_____	_____	_____	_____
Other Costs	_____	_____	_____	_____	_____	_____

Miscellaneous Costs Total _____

CAD/CAM APPROACH GRAND TOTAL _____

FIGURE 8-6 continued.

At this point, the cost projections for the two options can be compared. Figure 8-7 is a cost comparison worksheet that can be used for comparing the manual and CAD/CAM options for the specified period. Each data element contained in the worksheet would show the total cost estimated for that element for the specified period.

After comparing the financial forecasts, managers will have a good idea as to the feasibility of a CAD/CAM conversion. However, before arriving at a final decision it is well to examine some of the other factors

Projection Period _____

CAD/CAM OPTION   MANUAL OPTION

Personnel .... _____   Personnel .... _____

Initial hardware .... _____

Initial software .... _____   Equipment .... _____

Expansion .... _____

Training .... _____   Facility .... _____

Maintenance .... _____

Facility .... _____   Supplies .... _____

Supplies .... _____

Miscellaneous .... _____   Miscellaneous .... _____

CAD/CAM TOTAL .... _____   MANUAL TOTAL .... _____

FIGURE 8-7 Cost comparison worksheet for CAD/CAM versus manual approach.

which are indicators of the advisability of such a conversion. There are both positive and negative indicators. CAD/CAM students should be familiar with both.

Other Factors to Consider

Two other factors which can be considered in determining if a CAD/CAM conversion is justified are the *payback period* and *return on investment* (ROI). The payback period is the amount of time it takes a company to recover the initial investment. It can be calculated using the following formula:

$$\text{Payback Period} = \frac{\text{Adjusted Installation Cost}}{\text{Net Annual Savings}}$$

These terms are defined as follows:

Payback Period: The total amount of time that will elapse before a company will realize a positive cash flow from the CAD/CAM system.

Adjusted Installation: The total of all expenditures for the installation of the CAD/CAM system.

Net Annual Savings: The net annual savings attributed to the CAD/CAM system.

As an example of how this formula can be used, assume that the net annual savings for a given CAD/CAM system is $159,000 and the adjusted installation cost is $193,472. The payback period would be calculated as:

$$\frac{193,472}{159,000} = 1.22 \text{ years}$$

Once the payback period has been determined, it becomes a management decision as to whether it is acceptable.

The ROI can also be calculated and considered as a cost-justification factor. ROI tells managers if the return that will be generated by the investment in a CAD/CAM system is worth making the investment in the first place. The formula for calculating the ROI is:

$$ROI = \frac{\text{Average Yearly Return}}{\text{Adjusted Installation Cost}} \times 100$$

These terms are defined as follows:

Average Yearly Return: Average yearly savings generated by the CAD/CAM system.

Adjusted Installation Cost: The total of all expenditures for the installation of the CAD/CAM system.

As an example of how this formula can be used, assume that the adjusted installation cost for a CAD/CAM system is $193,472 and the average yearly return is $159,000. The ROI would be calculated as follows:

$$\frac{159,000}{193,472} \times 100 = 82.18\%$$

The ROI in this hypothetical case would be 82.18%, which is very favorable. An ROI of even 15%-20% is normally considered favorable.

If the projections accomplished in the previous section indicate that a CAD/CAM conversion is feasible, the decision should be to move forward. However, before doing so, managers are well advised to become familiar with some of the factors which can invalidate these projections as well as those which can improve on them.

Negative Factors

What follows is a list of factors which can have a negative effect on a CAD/CAM conversion. If one is aware of these factors, steps can be taken to overcome or at least minimize their impact.

- Downturns in the economy can invalidate the financial forecast upon which all projections were based.
- Rejection of CAD/CAM systems by key personnel.
- Failure of the vendor to comply with agreed on shipping, installation, and training schedules.
- A poor after-purchase relationship with the vendor.
- The vendor goes out of business.
- Hardware and software problems which cause continual workflow interruptions.

By understanding these factors and the impact they can have, managers can take all possible steps either to prevent them from occurring or to minimize their effects if they do.

Positive Factors

There are factors which can actually improve the savings projected in undertaking a CAD/CAM conversion. Managers should be familiar with these so as to be sure to promote such factors in any way possible.

- Better than projected increases in the workload because of upturns in the economy or other unpredictable reasons.
- Enthusiastic acceptance of CAD/CAM by key personnel.
- The vendor complies with or even exceeds all shipping, installation, and training deadlines.
- A positive after-purchase relationship develops between the buyer and the vendor.
- The vendor maintains fiscal soundness and continually improves its products and services.
- There are few workflow interruptions because of maintenance problems, and those which do occur are quickly and effectively dealt with.

EVALUATING AND SELECTING SYSTEMS

Having made a decision to undertake a CAD/CAM conversion, the next step is to select a vendor and the actual system(s). This is a critical phase in the overall CAD/CAM transition; possibly, the most critical phase. In making a CAD/CAM purchase, a company is buying more than hardware and software. It is also buying such intangibles as

1. the vendor's after-purchase philosophy,
2. the vendor's ability and commitment to stay in the CAD/CAM business,
3. The vendor's services, and
4. The vendor's knowledge of the application area in question.

Evaluating Vendors

Most people prefer to buy a car from a dealer with a long-standing track record, a good reputation, and a proven commitment to the after-purchase relationship. This is a wise approach. A similar approach is advisable with a CAD/CAM purchase. This section explains how to evaluate vendors, their services, their training, their after-purchase commitment, and other important intangibles.

In making an evaluation, the key is knowing what questions to ask and what criteria to use. The performance of a given CAD/CAM system is important. However, do not make the mistake of selecting a vendor solely on the weight of its system's performance. The integrity, credibility, dependability, and fiscal soundness of the vendor are at least as important as the performance of its system. Before purchasing a system from a given vendor, a buyer should take the time to get to know the vendor. What follows is an annotated list of criteria that can be used in evaluating vendors.

1. **What is the nature of feedback from other companies that have dealt with the vendor?** Ask the vendor to supply a list of other clients. Then contact people in these companies. Find out if the vendor is considered honest and dependable and if it has a stable group of people to deal with. It is difficult to build a positive relationship with vendor representatives if the representatives continually change because of turnover.

2. **Does the vendor have a written statement of its philosophy concerning the vendor/buyer relationship?** Know the official policy of the vendor so you have something to measure actual performance against. Show this written policy to other clients, and ask them to compare it to the actual performance they have experienced.

3. **How long has the vendor been in business, and how committed is it to staying in business?** It is important for the vendor to have a track record to judge. This is especially true for companies that are not familiar with CAD/CAM. A good rule of thumb is to let the more experienced users take their chances with new and unproven vendors. A well-established track record is an indication that the vendor plans to stay in business and is able to do so.

4. **Is the vendor fiscally sound?** The list of well-intended vendors that got into business with a good product but did not have the financial strength to make it is lengthy. No matter how good the system is, it will be of little use if the vendor that manufactures and markets it goes out of business. Consider the plight of individuals who purchased DeLorean cars. Before selecting a vendor, run a Dun and Bradstreet search on it. The vendor can also be asked to provide proof of its fiscal soundness. If this method is used, make sure that any data submitted are examined by competent, qualified financial experts.

5. **How does the vendor handle service and maintenance calls?** Is there a toll-free number for making service complaints? Does the vendor contract with local technicians or must they be transported from a central location? If they must be transported, what is the response time and how does it affect the cost?

6. **Does the vendor express an interest in working with potential buyers to clearly identify their needs to ensure that its system will meet those needs?** A reputable vendor will turn down a sale before it will sell to a buyer whose needs it cannot meet. Find out from the vendor what percentage of its total installations are in applications similar to yours. Even the best vendor in one area might be inexperienced in another area.

7. **Is the vendor willing to participate in side-by-side comparisons with other vendors?** Vendors make it their business to know the strengths and weaknesses of the competition. A great deal can be learned about vendors and their products in side-by-side comparisons with other vendors. Vendors who are not willing to participate in such comparisons should be examined even more carefully than usual.

8. **Does the vendor provide complete installation, startup, testing, and debugging services?** A vendor should provide complete services. Installation services should include at least the following:
   - Unpacking and set-up
   - Inventory of components to ensure that everything has been shipped
   - Hardware set-up and testing

- Software installation, trial runs, and debugging
- Establishing network interfaces
- Familiarization of the buyer with technical manuals
- Periodic spot-checking after installation

9. **How comprehensive are the training services provided?** The success of a conversion depends on the acceptance, skills, and creativity of people involved at all levels. These things will depend on the quality of the training provided. If managers receive the type of training needed, they will be able to handle the conversion in a way that will ensure its acceptance among employees. If employees receive the proper training, they will have the skills and creativity that are necessary to make a system productive. Vendor-provided training should include management training and training at the beginning and advanced levels for users.

Selecting Systems

The types of system(s) that might be included in a CAD/CAM conversion vary. There are CAD, CNC, and robotics systems. There are also CIM systems which might include all of these. Within each of these categories, there are many different makes and models of systems.

In spite of the diversity of systems available, there are common criteria which can be applied in making a decision to select for a given application in a given setting. These criteria include (Figure 8-8):

1. application factors,
2. operational factors,
3. performance factors, and
4. expansion factors

Application Factors

The main concern here is whether the system(s) can handle the application in question. This is the first and most important factor to consider. An otherwise excellent system that cannot handle the application in question should not be considered.

Operational Factors

How difficult is the system to use and how long does it take to learn how to use it? These are important questions. The system, regardless of type, should be "user friendly." This means it should be designed and configured for maximum operational convenience. If a system is difficult to operate, users will tire of it, productivity will decline, and the conversion will fail.

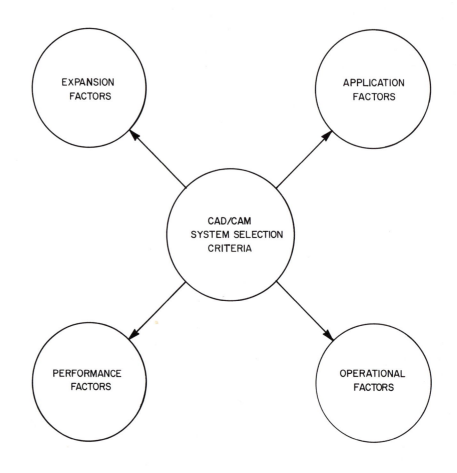

FIGURE 8-8 CAD/CAM selection criteria.

A system should be easy to learn. Because of the time required to learn a new system, productivity is likely to decline in the early stages of the CAD/CAM conversion. The easier a system is to learn, the shorter this period of decline is likely to be. It is important for users to be able to experience success and comfort with a new CAD/CAM system as soon as possible. The sooner they feel comfortable, the sooner they will accept the system. Correspondingly, a system that is difficult to learn is likely to be rejected.

Performance Factors

How fast and how well can the system do what it is supposed to do? These are important questions to ask because the answers will have an effect on the ultimate success of the CAD/CAM conversion.

"Throughput" is the term used to describe how fast a system can execute certain tasks. It is usually expressed in the number of executions per unit of time, such as executions per millisecond, second, minute, or hour. The throughput ratings for competing systems should be compared closely in selecting a system.

How well a system can do what it was designed to do can be measured a number of ways. Two commonly applied criteria for such measurements are accuracy and repeatability. Accuracy refers to how close the system can actually come to accomplishing the specified task. For example, if a CNC drill is supposed to bore a hole 1.50 inches from a given datum, how close to 1.50 can it locate the center of the hole? Repeatability is a measure of how often the machine can repeat such a task with the same degree of accuracy. Such ratings for competing systems should be compared closely.

A final performance criterion to consider is reliability. How much downtime is likely to occur due to system malfunctions and maintenance problems? Many vendors build features into their systems to minimize downtime. A list of all such features should be obtained for the various systems being considered. The lists can be compared. In addition to comparing these lists, it is well to ask other buyers of the system how much downtime they have experienced.

Expansion Factors

Can the system be expanded? How easily? At what cost? Expandability is an important consideration when selecting any type of CAD/CAM system. A CAD/CAM system should be easily expandable and at a reasonable cost. Some CAD/CAM systems are designed in modular form to promote expandability.

It is a good idea to require vendors to explain in detail what is required to expand a system. It is also a good idea to ask for cost estimates expressed in cost per unit or module. For example, the cost of expanding a CAD system can be given in cost per workstation. The cost of expanding a robotics system can be expressed in cost per robot. Cost per machine can be used for a CNC system.

PERSONNEL CONCERNS

There are a number of personnel related concerns of which CAD/CAM managers should be aware and with which they should be prepared to deal. These include staffing, training, and morale concerns. Managers who are aware of the various problems which can arise in each of these areas will be better prepared to prevent or at least deal with them.

Staffing Concerns

When undergoing a CAD/CAM conversion, managers must decide whether to employ experienced CAD/CAM personnel or to upgrade and retrain existing personnel. Both approaches can be successful and both can fail. Each has advantages and disadvantages.

The staffing needs which result from a CAD/CAM conversion fall into two broad categories:

1. Management personnel
2. Technical personnel

Management personnel plan, implement, and oversee the conversion. Technical people use the CAD/CAM systems that are installed.

Managers are responsible for such conversion tasks as

1. vendor and equipment selection,
2. installation planning,
3. training planning,
4. acceptance and sign-off of CAD/CAM systems after testing,
5. monitoring,
6. planning for system integration and networking, and
7. planning for future expansion.

Technical personnel are responsible for such conversion tasks as

1. testing of systems and making recommendations to managers about acceptance;
2. becoming proficient in the use of systems and helping train others; and
3. identifying hardware malfunctions and software bugs so they can be corrected by the vendor.

All of the tasks listed above must be performed in any CAD/CAM conversion. Clearly, experienced personnel will be better able to accomplish such tasks than managers and technical personnel to whom CAD/CAM is new. This is an important point to consider, because the ultimate success of a CAD/CAM conversion will depend on how well it is handled in the early stages. Because of this, many companies choose to employ experienced CAD/CAM personnel as part of their CAD/CAM conversion.

This can be a successful approach, but it can also fail. It is important to remember that the success of a conversion depends at least as much on people as it does on hardware and software. One problem that can arise when outsiders are brought in as part of a CAD/CAM conversion is rejection of them and the conversion by existing employees. This is especially true if existing employees are laid off to make room for the new ones.

Another problem can arise when the experienced CAD/CAM personnel who are brought in do not know the product or company. Knowledge of a given company and its products comes from inside experience. Once gained, such experience is difficult to replace. For these reasons, some companies choose to upgrade and retain their existing employees as part of the CAD/CAM conversion.

This can be a successful approach, but it can fail too. Such an approach will tend to lengthen the amount of time that will elapse before the advantages of CAD/CAM will be realized. This is the "learning curve" problem. Productivity, in the early stages of the conversion, will tend to go down rather than up with this approach.

Neither approach used exclusively is without shortcomings. For this reason, some companies choose a hybrid approach which combines the best of the two. With such an approach, existing employees are maintained to the greatest possible extent. If employees are laid off, it is not to make room for outsiders. Experienced CAD/CAM personnel are hired, but only as temporary consultants. With this approach, the strengths of the other two approaches are gained and the weaknesses eliminated.

In writing consultant contracts, it is a good idea to build in certain checks and balances. It is unwise to write an open-ended contract which keeps consultants on the payroll until the conversion is completed without setting target dates for when that might be.

A good approach is to write goals with accompanying timetables into the contract, with built-in incentives for beating deadlines and, correspondingly, disincentives for failing to meet them. This will shorten the amount of time in which consultants are needed without harming them financially. Regardless of which approach is taken, training will be one of the keys to a successful conversion.

Training Concerns

Training is a critical element in a CAD/CAM conversion. Managers and technical personnel should be trained as part of the conversion, but can experienced manual personnel learn CAD/CAM? Research has shown that they can.

Figure 8-9 is a pie-chart showing the results of research completed by Daratech Associates of Cambridge, Massachusetts. This graphic shows that 40% of manual personnel can become expert in the application of CAD/CAM. Another 40% can adapt well and become good users. Only 20% do not adapt well.

A concern when upgrading manual personnel for CAD/CAM is how long it will take them to become productive. Figure 8-10 shows the

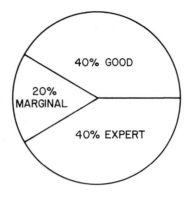

FIGURE 8-9 Percent of manual personnel who can successfully convert to CAD/CAM.

results of Daratech's research in this area. Within three months, manual personnel can be working at the intermediate skills level, and by nine months, they can be expert.

What are the types of topics which should be covered in a CAD/CAM training course? The content, of course, will differ for managers and technical personnel.

In a management training course, those topics which will help managers make the transition a success and manage the conversion once it is completed should be covered. They include the following:

1. Installation planning
2. Preventing and solving personnel problems
3. Realistic expectations
4. Facility adaptations
5. Planning training
6. System access
7. System management

In technical training courses, those topics which will help make people skilled, proficient users should be covered. They include

1. basic system operation,
2. intermediate techniques, and
3. advanced techniques.

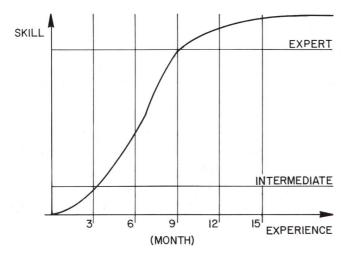

FIGURE 8-10 A typical CAD/CAM learning curve.

Morale Concerns

A CAD/CAM conversion can have a positive or negative effect on morale, depending on how it is handled. There are two key areas of concern:

1. Job security
2. Employee frustration

One of the ways that CAD/CAM conversions save money is by requiring less pesonnel. The savings in this regard can be significant, because personnel costs typically represent 80% of the fixed overhead costs in a traditional manufacturing plant. However, if not handled properly, what is gained in fiscal savings can be lost in morale problems.

Managers can minimize morale problems by providing retraining, job placement, and resettlement assistance to employees who are laid off as the result of a CAD/CAM conversion. When employees are laid off, there is normally a ripple effect among those who remain. An "it's you now, but will it be me later?" attitude is likely to develop.

Unaffected workers will watch to see how affected workers are treated. Unless they see their colleagues being treated fairly, they will tend to worry about their own job security, and morale will decline correspondingly. This is an important point to remember. Some managers have undermined the success of a CAD/CAM conversion by mistakenly thinking that the negative repercushions will be confined to those employees who are directly affected.

Employee frustration is another potential problem area. Working with computerized systems can be a frustrating experience, especially for people who are used to the traditional manual processes. A computer will do exactly what it is told, but only if properly instructed. Even the slightest deviation from required procedures will cause a CAD/CAM system to fail to respond in the desired way, and sometimes not to respond at all.

There are two precautions managers can take to minimize the number of problems that will result from worker frustration. First, managers should concentrate on selecting CAD/CAM systems that are user friendly, easy to learn, and simple to operate on a day-to-day basis. When upgrading manual personnel, the sooner they experience success on a system, the sooner they will begin to feel comfortable.

Second, managers should ensure that technical training for upgrading manual personnel is given careful attention. High-quality training will do more than anything to minimize the occurrences of worker frustration.

SUMMARY

It is important for students of CAD/CAM to know the technical aspects of CAD/CAM. It is also important for them to understand CAD/CAM from a management perspective. In attempting to determine if a CAD/CAM conversion is justifiable, managers should (1) identify colleagues in similar companies who have undergone a conversion and solicit their input, (2) analyze their own company's financial forcast, and (3) familiarize themselves with the factors affecting the advisability or nonadvisability of a conversion.

A major task to accomplish in attempting to decide whether a CAD/CAM conversion is warranted is the comparison of costs for the manual and CAD/CAM approaches. In estimating manual costs, managers should project personnel, facility, furniture and equipment, and expendable supply needs, as well as their corresponding costs.

In estimating the cost of the CAD/CAM approach, managers should project personnel, initial hardware, initial software, expansion, training, maintenance, facility, and miscellaneous needs, as well as their corresponding costs. After comparing the manual and CAD/CAM cost projections, managers should consider several other factors.

Factors that can have a negative effect on a CAD/CAM conversion and invalidate the earlier projections include (1) unexpected downturns in the economy, (2) rejection of CAD/CAM by key personnel, (3) failure of the vendor to comply with schedules, (4) a poor after-purchase relationship with the vendor, (5) a vendor's going out of business, and (6) work interruptions due to maintenance problems and bugs.

Factors that can have a positive effect on a CAD/CAM conversion and enhance projections include (1) better than projected upturns in the economy, (2) enthusiastic acceptance of CAD/CAM by key personnel, (3) the vendor's complying with or exceeding all agreed on schedules; (4) a positive after-purchase relationship evolving with the vendor, (5) the vendor's maintaining fiscal soundness, and (6) minimal workflow interruptions.

Evaluating vendors and selecting systems is a crucial step in a CAD/CAM conversion. In evaluating vendors, managers should ask a number of questions in such areas as satisfaction rate of other customers, after-purchase philosophy, commitment to staying in business, fiscal soundness, service and maintenance arrangements, willingness to participate in side-by-side comparisons, and services provided.

In selecting systems, managers should consider application, operation, performance, and expansion factors. Applications factors deal with the appropriateness of a given system for a specific application. Operational factors to consider are how easy the system is to operate and how long it takes to learn it. Performance factors include throughput, accuracy, and repeatability ratings. Expansion factors include how a system can be expanded and at what cost.

People will ultimately determine the success or failure of a CAD/CAM conversion. Consequently, managers should be aware of the personnel concerns which accompany a CAD/CAM conversion. They are staffing, training, and morale. Managers must decide whether to hire experienced CAD/CAM personnel and lay off the existing work force, upgrade the existing work force, or hire some specialists on a consultive basis and upgrade the existing work force wherever possible. Managers must ensure that both management and technical-oriented training are provided. In order to avoid morale problems managers should make sure that all personnel get the proper training, user friendly systems are purchased, and personnel who cannot make the transition are treated fairly and properly.

Chapter Eight REVIEW

1. What should a manager who is considering a CAD/CAM conversion look for when contacting other companies?
2. What four areas of cost should be estimated in projecting the cost of the manual approach?

3. What eight areas of cost should be estimated in projecting the cost of the CAD/CAM approach?
4. List five factors that can invalidate the projections made in justifying a CAD/CAM conversion.
5. List five factors that can enhance the projections made in justifying a CAD/CAM conversion.
6. Explain how you would select a vendor when considering a CAD/CAM conversion.
7. Explain the four criteria that should be applied in choosing a specific CAD/CAM system.
8. Explain the staffing concerns facing managers undertaking a CAD/CAM conversion.
9. Explain the training concerns facing managers undertaking a CAD/CAM conversion.
10. Explain the morale concerns facing managers undertaking a CAD/CAM conversion.

All of the well-known, major CAD/CAM concepts have been discussed. They include CADD, CNC, robotics, and CIM. When the term CAD/CAM is used, these are usually the concepts being spoken of. However, CAD/CAM is a broad term which encompasses more than just CADD, CNC, robotics, and CIM. In reality, it encompasses any use of the computer as an aid in design, drafting, and manufacturing.

In addition to the well-known CAD/CAM concepts, there are several lesser known concepts with which students should be familiar. These include Manufacturing Resource Planning (MRP), computer-aided maintenance, Computer-Aided Engineering (CAE), and computer-aided quality control/inspection (Figure 9-1, page 293). All of these are concepts in which the computer is used as an *aid* in performing design, drafting, or manufacturing-related jobs or tasks.

Major Topics Covered

- Manufacturing Resource Planning
- Computer-Aided Maintenance
- Computer-Aided Engineering (CAE)
- Computer-Aided Quality Control and Inspection

Chapter Nine

Other CAD/CAM Concepts

The CAM Systems group at Metcut Research has developed an integrated process planning system for rotational parts for the Aircraft Engine Group of the General Electric Company. The system utilizes group technology as a primary technique to organize and process the manufacturing data in the process planning data base. Functions such as cost estimation, development of routing sheets, selection of tooling and feeds and speeds, graphical verification of NC tapes, and graphical composition of operation instructions can be performed by the system. Hardware to support the system includes a Perkin-Elmer minicomputer, a refresh graphics CRT, a digitizing tablet, and a plotter. This paper will describe the capabilities of the system and present actual computer outputs.

INTRODUCTION

The General Electric Aircraft Engine Group (AEG) is pursuing an aggressive program to implement and integrate CAD/CAM in the design and production of aircraft engine components. The overall CAD/CAM program incorporates several interactive graphics systems, NC programming systems, automated quality control, automated factory management, process planning, and an extensive DNC network with graphics capabilities at the workstations. The process planning system is being developed by Metcut Research Associates Inc. This paper will present the development and current status of the process planning system.

BACKGROUND

General Electric was one of the first companies to use interactive graphics in the design of production parts. The CAD activity at General Electric/AEG was well-founded during the early 1970s. In the CAM area, General Electric was among the original users of NC equipment. In 1975, a

Courtesy of SME, "Integrated Process Planning at General Electric's Aircraft Engine Group," by Steven A. Vogel and Diane Dawson.

major Master CAM Plan for the Rotating Parts Operation (RPO) was developed to expand the CAM modules and provide a closer integration with the CAD activities.

The RPO CAM Master Plan integrated existing CAM modules and provided for the development of additional CAM capabilities. The overall CAM structure incorporates DNC, quality control, NC programming, factory management, machine tool monitors, and process planning.

PROCESS PLANNING

Metcut Research initiated a three-phased approach to the development and implementation of the process planning module for the RPO CAM Master Plan. During Phase I, which was performed in 1977, a system design was developed. An extensive analysis of the manual process planning function was undertaken during this phase to establish the status of process planning at that time. Based on this analysis, a design was for a computer-aided process planning system which automated the current process planning procedures. A key consideration in the Phase I design was the development of effective interfaces with other computer data bases that supply data to the process planners.

The Phase II activity was performed in 1979 and 1980. During this phase, the system design was programmed and implemented on a prototype basis. A subset of rotating parts was selected as the pilot parts spectrum and the system was fully developed for turning operations. A pilot computer workstation was installed in the RPO process planning area to introduce the system into production. All of the modules of the process planning system are operational in the prototype system.

The Phase III activity is currently being performed. It is scheduled to be completed by the end of 1981. During this phase, the system will be expanded to cover the process planning activities for all RPO parts and all types of operations.

(continued)

BASIC APPROACH

The RPO manufactures aircraft engine parts that are designed with close dimensional and surface finish requirements. The manufacturing sequence may include over one hundred distinct manufacturing steps on different machine tools. In fact, the parts sometimes travel to outside vendors for certain operations, then are completed in-house. The complex designs of the parts and the exotic workpiece materials often necessitate specially designed tooling and complicated manufacturing techniques that have evolved over the course of many years of manufacturing experience.

Because of this, the basic philosophy behind the process planning system is not to introduce a "computerized" process planning system which is capable of automatically making all of the decisions necessary to process a part. Rather, it is a "computer-aided" system which provides powerful capabilities in group technology retrieval, computer graphics, generative process planning, and process optimization. These computer tools can be used by a process planner to complement his intuition and experience to arrive at cost-effective process plans.

Another fundamental concept that is key to the process planning system is the development of effective interfaces with other CAD/CAM databases. Many CAD/CAM modules with their own databases have been established at General Electric in the past ten years. Many of these systems provide input that is useful to the computer programs that compose the process planning system. Interfaces were established to get data from the CAD databases, tooling databases, and DNC databases that already exist on various computers in the plant. Interfaces were also established to send output from the process planning database back to systems which require process planning information.

GROUP TECHNOLOGY RETRIEVAL

A problem that is common to any large manufacturing organization is the difficulty in exchanging manufacturing information between various process planners and between manufacturing departments. This problem is aggravated in manufacturing departments that process a large number of parts and have sophisticated manufacturing techniques.

The RPO process planning system incorporates group technology retrieval capabilities that address the problems of data retrieval. Specialized coding schemes have been devised for the part information and manufacturing information that must be shared among process planners and other departments. Three coding schemes have been developed. The first is a group technology code for part designs. The second scheme describes tooling designs. The third group technology scheme is for process descriptions. By specifying values for these coding schemes in combination with one another, it is possible to retrieve historical processing information for current process planning efforts. For example, forging designs that have been used for a certain part configuration can be retrieved to locate existing forgings which have potential applications on new parts. Similarly, it is possible to retrieve candidate tooling designs by specifying the important geometric and technological requirements of the tooling application. This ability to retrieve tool designs is useful to reuse "special" tooling designs on new parts rather than reinventing identical or similar tooling for each application.

Processing information can be retrieved via the group technology scheme for process description. For example, entire tooling layouts can be retrieved by operation type. Other factors of production such as feeds and speeds, work instructions, routings, quality plans, and so forth can also be retrieved in this manner.

The ability to retrieve historical data is directly useful to "share" historical planning information. It also provides a means to standardize manufacturing methods and to stop the proliferation of the variety that is common to most manufacturing areas.

(continued)

COMPUTER GRAPHICS

One of the more "eye catching" aspects of the RPO process planning system is the extensive use of computer graphics throughout the entire process planning cycle. A fundamental decision was made in the design of the system to extend the use of tooling and part models from CAD systems to process planning activities. The graphical aids for process planning fall into four categories: (1) ability to overlay part designs and forging envelopes, (2) ability to create graphical tooling layouts, (3) NC machining simulation, and (4) development of graphical work instructions.

The capability to overlay part designs and forging envelopes is useful in conjunction with the group technology routines to evaluate historical forging designs. Candidate forgings can be located through a group technology search and overlaid on top of a new part design. This will give a graphical check of the suitability of existing forging designs for new parts. It also is useful in analyzing the parts spectrum to determine geometric variations, thereby providing a tool for standardization.

Because the RPO manufactures rotational parts, the major machining operation is turning. There are a large number of turret lathes which are set up with a variety of turret configurations. Each turret configuration has tool blocks, tool holders, and inserts on as many as six faces of the turret. Historically, these turret configurations have been drafted by hand for each turning operation.

One of the graphical aids available in the process planning system is a graphics package which interactively composes turret layouts on the CRT. Starting with a bare turret, predefined mating surfaces on each tool item can be mated together with tool holders, adapter plates, tool blocks, and inserts. The resulting turret configuration can be plotted, used for NC simulation, and used in the graphical work instruction program.

NC simulation in the process planning system is more realistic and thorough than in most CAD/CAM systems. Because the entire turret layout is predefined, it is possible to drive the entire tooling configuration through a cutter path with the part, forging, and fixture simultaneously displayed on the CRT. Not only does this approach show errors in the path of the cutting tool, but it also provides interference checking between other tooling on the turret and part, forging and fixture.

Another powerful graphical aid for the process planner is an interactive documentation package. Using this program, a planner can call up graphical images of part, forging, tool layout, fixtures, and in-process workpiece outlines to compose graphical work instructions for the machine tool operator. Simple commands allow the planner to compose the picture by dragging the graphical items to positions on the screen, adding text, lines, arrows and symbols, and plotting the resulting display or storing it for later retrieval. Ultimately, the graphical work instructions will be sent directly to a graphics terminal at the machine tool via a DNC line.

GENERATIVE PROCESS PLANNING

The group technology codes for part designs, tool designs, and process descriptions provide a way of storing technological data that can be used to support generative process planning. Decision logic has been developed and is continuing to be developed to turn this technological data into manufacturing decisions. This technique is being used to automatically define operation sequences, machine tool selection, tooling selection, cut selection, and feed/speed selection.

The decision logic itself is stored in data files to allow flexibility as manufacturing methods change. The logic can be developed by the most experienced planners so that the best, most current manufacturing decisions are reflected in the decision logic. Nonetheless, the generative capabilities can be "overridden" by the process planner if the generated processing methods were not acceptable.

(continued)

PROCESS OPTIMIZATION

The RPO process planning system takes advantage of the latest technological approach in optimizing process parameters based on minimum cost and maximum production rate. By evaluating the trade-offs between higher metal removal rates and higher tool costs, productivity can be increased by as much as 30% and costs reduced by a corresponding figure.

The process optimization capability requires data from several sources. A machinability database, which is a subset of the process planning database, includes mathematical models of tool life, cutting forces, surface finish, and horsepower as a function of feed, speed, and depth of cut. Economic information pertaining to cutting tools and machine tools are stored in other files in the process planning database. A program which evaluates cutter/workpiece overlap calculates cut geometry for all NC cuts based on the original forging outline, the tooling configuration, and the NC cutter path.

Economic equations have been devised which utilize all of this information to select optimum feeds and speeds. The computer first selects a qualified region of feeds and speeds based on surface finish, cutting force, and horsepower constraints. Next, the machining cost and time is calculated at positions within this constrained region to arrive at the optimum cutting conditions. These conditions are then programmed into the NC tape.

JUSTIFICATION

The major impact of the Computer Aided Process Planning System is the utilization of the computer to perform tasks that were previously done manually, and to identify methods of production which are more cost effective than existing methods. The most promising areas for cost savings are in the graphic tool layout and verification capabilities, and in optimization of the manufacturing process.

The graphical turret layout capability supplies the planner with the same information as drawings previously supplied by the Tool Design Component. The accessibility of existing designs through the use of group technology techniques will also tend to standardize layouts and reduce special tool designs. Consequently, these techniques will greatly reduce the manual tool design effort associated with the implementation of any new process plan.

Verification of new NC tapes will be improved by the visual technique of driving the tooling configuration through the cutter path, by designating cutting force constraints, and by confirming the appropriateness of tooling selection using the mathematical models available. These will reduce interference and breakage problems on NC tape tryout and reduce the associated cost of damaged tooling and parts. The primary benefit of the system is its process optimization capabilities. The additional planning time made available through this system by reduction of the tedious manual work associated with planning activities will be available to optimize existing processes. Even though process optimization can be performed manually, it was not done because of the extreme amount of time required for the manual calculations. It is estimated that a 5% productivity improvement will be attained annually for several years with the optimization of process parameters.

The impact of the computer-aided process planning system on the Rotating Parts Operation is estimated to be on the order of $1,000,000 annually when it is implemented throughout the shop. As familiarity with the system capabilities increases, further applications and additional benefits are anticipated.

CONCLUSIONS

The implementation of this system has proved to be a most rewarding experience. The system has proved its worth in terms of reduced

(continued)

cost and improved productivity. Process Planning is one of the key elements of the integrated CAD/CAM system envisioned in the future of General Electric production. Furthermore, this pilot system work has indicated the potential for even greater technical innovations and has provided some of the tools required to advance the process planning system into a more powerful and cost effective component of an integrated CAD/CAM system.

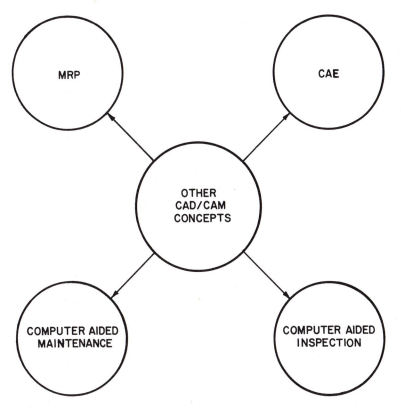

FIGURE 9-1 Other important CAD/CAM concepts.

MANUFACTURING RESOURCE PLANNING

Manufacturing Resource Planning (MRP) is a system which ties together capacity requirements planning, production planning and scheduling, purchasing, inventory management, and several other processes to ensure optimum productivity, on-time orders, the lowest possible work-in-process inventory, and the highest possible inventory turnover. The computer has become an invaluable aid in MRP. Some of the most important elements of an MRP system are production scheduling, process planning, purchasing, and sales.

Production Scheduling

Production scheduling is a common manufacturing task that can be simplified with the aid of a computer. Group technology has simplified production scheduling, as you already know (Chapter 6). Group technology coupled with the computer can simplify the process even more. An example will illustrate this point.

Prior to group technology, the number of possible sequences for producing parts could be exceedingly high. The number of possible sequences for running X parts over Y machines can be calculated as follows: X factorial to the Y power (X^Y). The number of possible sequences for running just five jobs across five machines would be:

$$(1 \times 2 \times 3 \times 4 \times 5) = 120$$

This is a mind boggling number of possible sequences. To glean from these the optimum sequence would be a formidable task if undertaken manually.

Because group technology removes one random factor, it reduces the number of possible sequences to five factorial (5!). However, this is still a lot of possible sequences (120) to deal with. If the number of jobs is six, the possible sequences increases to 720. If the number is seven, the possible sequences increase to 5,040. Even with group technology in place, selecting the optimum sequence from among so many is difficult. This is where the computer becomes an important aid in the process.

Performing the computations necessary to determine the optimum choice from among a large number of possibilities is a strength of the computer. Such tasks can be performed in minutes. Because of this, the computer has become an invaluable aid in production scheduling.

Process Planning

Estimation and process planning are time-consuming, difficult, and error prone when undertaken using the traditional manual methods. Both processes can be simplified with the aid of a computer. The task of the estimator is to determine as closely as possible how much it will cost to produce a product. This means he or she must estimate all costs, including labor, material, overhead, and hidden costs, so that the profit margin can be added to the cost. Accurate estimating is essential to success in manufacturing.

Process planning involves layout, tool design, and setup. This process, too, is difficult, time-consuming, and error prone when accomplished using traditional methods. Like estimation, process planning can be enhanced with the aid of a computer.

Layout involves developing a plan for meeting the tolerance, finish, feature, and dimensioning requirements of a job. Most errors in layout are caused by insufficient time to perform all of the calculations necessary to determine the optimum cycle time, mathematical mistakes, insufficient allowances for secondary operations, and mistakes in reading drawings.

Tool design involves a variety of mathematical calculations, looking up figures in tables, and making drawings; all of which are error-prone tasks when accomplished manually. Setup done manually is also an error-prone process. It involves preparing specific detailed instructions. Such instructions must cover gage point locations, what tools are needed, and the physical identification of all fixtures, gears, cams, and attachments.

The computer is a particularly useful tool for simplifying and streamlining both estimating and process planning. There are a number of cost-estimating software packages on the market which contain all of the estimating parameters needed in a manufacturing setting. With such a program, an estimator need only enter the appropriate information as prompted by the computer. The computer performs all calculations, research, and retrieval tasks required. The results of every estimate performed are categorized, stored, and updated immediately. With such a program, any person who is able to respond to the prompts can receive an accurate estimate in a matter of minutes.

Interactive graphics software is available for layout, tool design, and setup. Again, in each case, the computer and the program do the work. The operator simply responds to prompts. The advantage of the graphic capability is that it allows manufacturing personnel to *SEE* the impact of their responses.

Purchasing

Purchasing is an important element in the overall production process. All of the materials, spare parts, and components needed in manufacturing must be purchased at some point. The more efficient its purchasing operation, the more efficient a manufacturing firm will be and the less costly the inventory it will have to maintain in-house.

The computer can streamline the purchasing process. Several vendors produce software that allows manufacturing firms to network their purchasing office with its most frequent suppliers. In this way, items can be selected and ordered by computer. This is more efficient than the old method of searching through volumes of catalogs, filling out requisitions, and preparing purchase orders. It can even cut out the step of calling several companies for quotes.

With a computer network, the desired item can be called up, displayed on the screen along with pertinent data, and priced. The same process can be completed for several vendors of the same item. Once an order is placed by a manufacturer, the vendor's computer processes it, updates its own inventory, prepares the shipment, and writes the invoice.

Sales

Two problems frequently faced by industrial sales people are (1) insufficient knowledge of items in stock and (2) insufficient in-depth knowledge of the product line. Both of these problems can be solved with the aid of the computer.

When a salesperson is making a call, he or she needs accurate, up-to-date information concerning the availability of items a customer might order. It is poor practice to accept an order for more items than exist in stock without notifying the customer immediately of the need for a back order and how long it will take to fill it. Traditionally, this has been a common occurrence because sales personnel working in the field have to rely on printouts that could be outdated upon receipt.

Many distributors have solved this problem with the aid of a computer. By placing their inventories and ordering procedures on-line in a centralized computer and giving sales personnel access to the mainframe via briefcase-size microcomputers and a modem, the problem can be solved. When a customer wishes to order a certain quantity of a given item, the salesperson simply calls up the home computer and gets a readout on what is available. If there is a sufficient supply in stock, the order is placed immediately by computer. Some systems allow salespeople to "lock in" items in this way pending receipt of an order. Others allow their salespeople to actually place the order on-line.

Computers can also be used to solve the knowledge gap problem. Although most sales people have a good working knowledge of the product line, they don't always have the in-depth knowledge needed to help a customer select the right item from among several options. Software is available which allows the sales person to make the right choice by entering specifications into the computer in response to prompts. Once all prompts have been answered, the computer makes the choice.

COMPUTER-AIDED MAINTENANCE

Equipment maintenance is a problem in modern manufacturing plants that can be solved with the aid of a computer. Computerized maintenance management systems are more efficient than the traditional manual systems, and they offer the same advantages typically associated with computers: (1) reduced labor, (2) reduced costs, and (3) improved productivity.

Maintenance management systems typically have three components (Figure 9-2), all of which lend themselves to computerization:

1. Equipment history component
2. Preventive maintenance component
3. Spare parts inventory

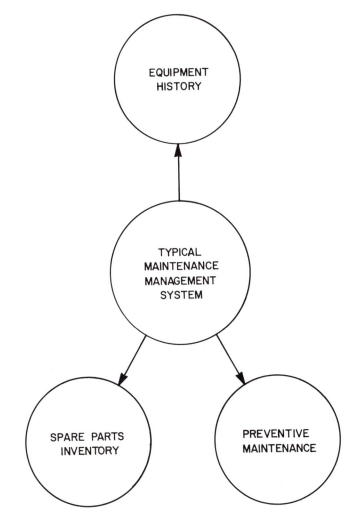

FIGURE 9-2 A typical maintenance management system.

Equipment History Component

The computer makes it especially easy to record and monitor equipment history records. Once stored, such records may be recalled, displayed, and interacted with in a matter of seconds. Information that might be included in a historical record for a piece of equipment includes the following:

1. Model name, manufacturer, and number
2. Equipment description
3. Purchase date
4. Installation date
5. Expected life span
6. Maintenance and repair history

Once such a file is established for each piece of equipment, keeping it current and interacting with it for any other reason is a simple, efficient process.

Preventive Maintenance

Using a computer, a preventive maintenance schedule can be developed for all equipment. Such a file should contain the model name and number for each piece of equipment, followed by a list of preventive maintenance procedures to be accomplished. Each procedure is assigned an initialization date and an explanation of how often thereafter the procedure is to be accomplished. With such a component in place, the maintenance engineer can begin each week with a printout of all maintenance tasks to be accomplished on what machines.

Spare Parts Inventory Component

Keeping track of spare parts has always been a problem for maintenance personnel. This is a problem that can be solved with the aid of a computer. By setting up a spare parts inventory in the computer, maintenance personnel can easily keep track of what is in stock, what has been used, when, in what quantities, and what should be reordered.

Each spare part item is listed along with a corresponding reorder number. A warning mechanism is triggered in the computer each time the inventory falls below the reorder number for a given part. Each time a part is taken from inventory it is entered into the computer and, in turn, subtracted from the inventory. When the reorder number is reached, the warning mechanism is triggered, showing what part is in low supply and how many to order.

As progress toward the completely automated factory continues, the computer will become the most widely used tool in manufacturing. The applications explained herein are only a few of the many that will be seen in the future.

COMPUTER-AIDED QUALITY CONTROL AND INSPECTION

Quality control became an integral part of manufacturing with the advent of work specialization, and it still is. Two major elements of quality control are inspection and testing. The computer can be used to enhance both processes.

Inspection and testing have traditionally been accomplished manually, using a variety of gages and special devices. In a computer-aided quality control setting, they are accomplished using a computer and various sensor devices.

Using a computer and contact as well as non-contact sensors, all of the various characteristics of a part (i.e., shape, length, width, thickness, feature locations, parallelism, perpendicularity, flatness, circularity, profile, concentricity, symmetry, etc.) can be tested much faster than with manual methods.

Computer-controlled contact sensors include mechanical probes and coordinate measuring machines (Figure 9-3). Computer-controlled non-contact sensors fall into two broad categories: optical and non-optical. Optical devices and methods (Figure 9-4) include machine vision, laser scanners, and photogrammetry. Non-optical devices and methods (Figure 9-5) include ultrasonic devices, radiation, and electrical energy methods. As technology in this area continues to evolve, the non-contact sensors will eventually become the norm.

COMPUTER-AIDED ENGINEERING

Engineering involves putting scientific knowledge, principles, and theories to practical use. In other words, engineering is the application of these things. Consequently, computer-aided engineering is using a computer and peripheral devices to aid in the application of scientific principles to the solution of practical problems. Computer-Aided Engineering is a broad term that encompasses several that have already been covered, including CAD, CAM, and CNC.

Computer-Aided Engineering, or CAE, is a term sometimes used to collectively describe all of the various processes which take place between the time when an idea is arrived at and that idea becomes a product.

Those processes involve design, manufacturing, testing, and all of the interactive subprocesses that go with them. The computer can be an invaluable tool in all of these processes.

The broad, all-encompassing definition of CAE is not the one which has caught on in the language of engineering. A slightly more

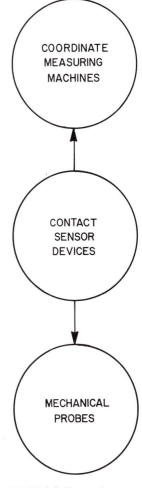

FIGURE 9-3 Types of contact sensor devices.

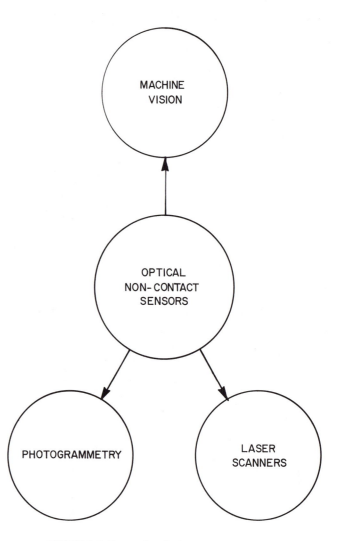

FIGURE 9-4 Types of optical non-contact sensors.

restrictive definition which limits CAE to just the engineering aspects of those processes which take place between the extremes of the idea and the finished product has become the more widely accepted version.

In this version, all of the various elements of the overall engineering process are referenced and included. These elements are

1. systems engineering,
2. design engineering,
3. manufacturing engineering, and
4. test engineering.

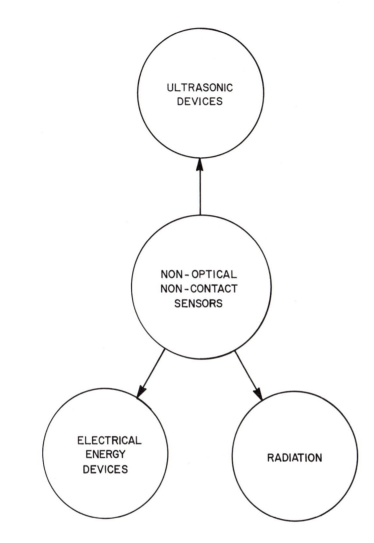

FIGURE 9-5 Types of non-optical, non-contact sensors.

CAE is coming to be accepted as a way of tying all of these elements together with one common database. In addition, any use of the computer to enhance the capabilities of persons working in these elements is considered CAE.

SUMMARY

Manufacturing resource planning is a computer-dependent system which ties together such manufacturing components as capacity requirements planning, production planning and scheduling, purchasing, inventory management, sales, and accounting to ensure optimum productivity, on-time orders, the lowest possible work-in-process inventory, and the highest possible inventory turnover.

Computer-aided maintenance involves using a computer to enhance the operation of a company's maintenance system. Such a system will normally have three major components: equipment history, preventive maintenance, and spare parts inventory. Each of these components can be automated using a computer-aided maintenance system.

Computer-aided quality control and inspection involves using a computer and a variety of peripheral devices to enhance the quality assurance processes commonly applied in manufacturing plants. All characteristics of a manufactured product such as shape, length, width, thickness, feature locations, and so on can be inspected using a computer and contact or non-contact sensors.

Computer-Aided Engineering (CAE) is sometimes used as a broad term that encompasses all of the processes involved in transforming an idea into a finished product with the aid of a computer. A more widely accepted definition, however, limits CAE to using a computer to enhance the specific engineering aspects of these processes, including systems engineering, design engineering, manufacturing engineering, and testing engineering.

Chapter Nine REVIEW

1. Explain how the computer has become an invaluable aid for each of the following elements of manufacturing resources planning:
   Production scheduling
   Process planning
   Purchasing
2. Explain each of the following components of a computer-aided maintenance system:
   Equipment history component
   Preventive maintenance component
   Spare parts inventory component
3. What types of characteristics of a part can be inspected using a computer and contact or non-contact sensors?
4. Name two contact sensor devices.
5. Name two non-contact sensor devices.
6. Define the term Computer-Aided Engineering.

CAD/CAM has given the manufacturing industry of this country a badly needed lift. Until the advent of the various concepts which fall under the CAD/CAM umbrella, the manufacturing industry was showing only slow growth at best. CAD/CAM has begun to turn this situation around. As CAD/CAM technology continues to develop, the manufacturing industry should become increasingly revitalized. This chapter covers some of the major future developments which can be expected in CAD/CAM.

Major Topics Covered

- Future Developments in CADD
- Future Developments in CNC
- Future Developments in Robotics
- Future Developments in CIM

Chapter Ten

CAD/CAM and the Future

PROJECT BACKGROUND

This case study focuses on an industrial experiment conducted at the Holset Turbocharger Plant at Madison, Indiana. The line being upgraded manufactures the cast iron bearing housings which support the turbocharger shaft.

At the onset on this project, all equipment were mainly stand-alone type with no communication nor material flow via automation. With the exception of the honing and washing operations, all loading and unloading were performed manually. Manufacturing data collection and transmission were nonexistent. It was then decided that such a manufacturing line should be considered as a candidate of experimental integration.

SCOPE AND OBJECTIVES

Holset is a wholly owned subsidiary of Cummins, who has been engaged in manufacturing research partnership with Purdue University via the CIDMAC Program. Such relationship enabled student participation in an industrial experiment. The CIDMAC researchers met with Holset's engineers and jointly designed the following experiment:

Goal: To create an integrated manufacturing cell using existing equipment (with minimum addition of new equipment). Upon completion, material flow and manufacturing data transmission between stations will be automatic. As such, the completed cell will be capable of being linked to the factory's LAN.

Objectives:

1. An automatic data collection system.
2. Automatic Q/C process.
3. Tool change optimization.
4. Preventive Maintenance Scheduling.
5. Downtime Accounting.

Courtesy of SME Technical Paper #MS86-344 1986 "From Stand Alone to CIM" by Kwok-Sang Chui and A. Vandenberghe.

At this time, the SWT lathes have been linked to the robot-Kingsbury station by the new power conveyor. Yet from the control standpoint, all machine tools and their PCs are still operating on a stand-alone basis. The experiment will be conducted under the basic assumption that production disruption will not be tolerated, and that the project will not provide any relief to the production bottleneck which is controlled by the machine capacity of the SWT lathes.

DESIGN OF THE INFORMATION SYSTEM

The information system is designed to handle three main levels of data hierarchy, namely, the management level information, the status level information, and the control level information.

The management level information will deal strictly with performance and downtime-related statistics. Status level information, on the other hand, will handle the statistics which will be required for Q/C, P/M, and tool policy programs. These two levels of information are less time-critical as compared to the control level information. The control level will in part deal with the Data Highway protocol, and partially will deal with the actual machine control of each operation station. Specifically, the structure of these three levels of information includes the following:

Management Level
- Capacity vs. Results—Number of parts produced per time period.
- Repair Statistics such as causes of breakdown, time waiting for repairs, and actual repair time, etc.
- Scheduled downtime for P/M Program (how close to target).

Status Level
- Adjustment of P/M schedules
- Inspection Process—1st piece inspection, final inspection, and/or in between operations inspections.
- Schedules for tool-change policies

(continued)

- Automatic Q/C data collection
- Other production status related information

Control Level

- Most time critical as it deals with the communication between control, production, and interface devices through the I/O racks.
- Data reservoir for the other two levels of information.
- Overall operation surveying and triggers corrective action when needed.

Much effort has been expended to ensure the design of the control level information being capable of supporting the other levels of the hierarchy, while at the same time allowing smooth data transmission between operating stations. Hardware and software implementation will be outlined in the next section.

PROJECT IMPLEMENTATION

A key nontechnical activity needs to be mentioned prior to the discussion of project implementation. At the onset of this experiment, a joint meeting between manufacturing plant, the central support group, and the CIDMAC research group was held to lay the ground rules and to set direction for the experiment. Such activity later proved to be crucial to the success of the project. Because of this nontechnical activity, the project was well defined and supported by all parties involved. Such group consensus thus far has yielded handsome dividends in terms of organizational and technical supports. Also, unnecessary roundabouts were eliminated, which resulted in significant cost and time optimization.

Machine-Tools/Equipment Linkage

The two SWT CNC Lathes were kinked to the Kingsbury drilling station by way of a power conveyor, thus they completed the material flow cycle and eliminated any use of forklifts to move parts from one station to another. At the completion of the conveyor, one operator can handle the whole line, including manual loading and unloading of parts at the starting and finishing points of the line. Two key points need to be noted here. First, in spite of a continuous flow of material, each machine and its controller still operated at semi-stand-alone mode. There was no communication at this point between stations, and bad parts could not be identified until final inspection. Secondly, data for Q/C and P/M programs could not be collected automatically.

Information System-Hardware Specifications

The existing Allen-Bradley PLC-3 will be used as the central processor for the information system. New Data Highway modules and a peripheral communication module 1775 GA will be added to complete the system. The Modicon 484 will either be replaced by an AB PLC or removed from operation. In the process of hardware evaluation, four alternatives were investigated. The technological content of each alternative varied in direct proportion to the capital requirement.

Alternative One is a bare bones alternative which could suffice for the present status of the project. No Data Highway would yet be implemented. Communication would only be with the AB PLC 2/30 on the conveyor system through the ASCII module. However, the ASCII module was developed as a first step in the direction of data acquisition from a programmable controller and has only limited capabilities in doing so. Furthermore, the function of the module would become obsolete as soon as Data Highway is implemented.

Alternative Two consists of the underlying design used in Alternatives Three and Four: Data Highway as a medium for information flow. However, no extra help would be included from a GA module or software package. An IBM PC would be used as a "smart" terminal again. Extensive

(continued)

programming would be required both on a high level (BASIC at the IBM PC) and on a low level (ladder logic at the programmable controller). This alternative reflects the idea that information transfer be handled with only the bare essentials in terms of hardware. So if this hardware cannot solve a particular problem, the user will have to develop his/her own software to do so.

Alternative Three makes use of a software package called "Optimum Network Executive" by Tele-Denken Resources Inc. The software allows for access to memory of the programmable controllers through a high-level language (Assembley, BASIC, or C) and is specifically designed for use with Data Highway and an IBM PC (XT or AT). Programming is done on a higher level than the time-consuming ladder logic; however, the user will still have to provide application programs for protocol handling. Different software packages might perform memory access in various ways but will usually not eliminate the need for application programs as the GA module does.

Alternative Four does not need a "smart" terminal such as an IBM PC. The 1775 GA module contains a microprocessor (68000) which is parallel to an IBM PC in terms of processing capabilities. The module has its own high-level language (GA BASIC) and virtually eliminates the need for application programs for protocol handling (this is actually done internally in the module). Also, each individual processor on the line (the AB 2/30s) will not have to bother with executing code for sending out information. The GA module is capable of reading the memory of any of the processors on the Data Highway. Troubleshooting of ladder logic by the electricians should also be less bothersome, since programming in GA BASIC will prevent the ladder logic from getting cluttered up with information processing logic.

The decision was made, based on future application/potentials, to use the DATA Highway approach. Furthermore, the GA module in Alternative Four will fit well in the environment of the programmable controllers; it is capable of supplying the users with the speed and efficiency of a high-level language.

INFORMATION SYSTEM—SOFTWARE DESIGN

Two key programs are needed to trigger the Data Highway into functional mode. The first of those two pieces is for the "smart terminal, an IBM XT. This program has been written in "BASIC." For reference purposes, the IBM XT Program has been split up into nine sections:

Section 1. opens up the communication port and initiates all arrays that will be used;

Section 2. puts the first menu on the screen; checks the keyboard and communication buffer simultaneously;

Section 3. checks the flags which indicate the status and reasons for machines in down mode, provides the operator with "alert" message;

Section 4. provides operator with statistics of either production status or tooling condition; also, in each case allows the operator the avenue to return to the main menu;

Section 5. details of production statistics such as tool number, part count, maximum tool count, and residual tool limits;

Section 6. details of downtime or tooling condition resulting from tool failure, unscheduled tool change, or other reasons;

Section 7. allows operator to reset or change tool-life limits and it's closeness value; correct password will be required to enter such function;

Section 8. displays the tools that should be replaced while machine is in down mode; and

Section 9. consists of three subroutines, namely, the "clock reset," the "current count," and the "closeness compared"; in addition, Section 9 also contains the various routines called by the main program.

(continued)

The second piece of the key programs is the Allen-Bradley GA Module Procedure. The logic for the modules is designed to be in GA BASIC. In flowchart form, such logic consists of four procedures. Briefly, the first procedure deals with the Data Highway, and in a loop reads and files tool count information from all the stations. The second procedure reads the tool counts from the Data Highway and sets the GA count limit. The third procedure writes the accumulated values of the Data Highway which will be sent to the RS232C port of the IBM PC. The fourth procedure on the GA module acts as receiver of information over the RS232C port. To be more specific, the fourth procedure is responsible for sending information to the IBM terminal upon request of a flag. The logic of all these procedures has been designed, but details of programming are yet to be completed.

CURRENT STATUS AND FUTURE PROGRESS

Integration of machine tools has been completed, thus allowing material flow from start to finish requiring little or no manual support. Initial loading, and the unloading of finished parts, are on manual mode but can be mechanized easily when the need arises. The architecture of the overall Data Highway system has also been completed, from the logic design standpoint. Program for the IBM XT terminal has been written in BASIC. The Allen-Bradley module procedures are yet to be coded. Hardware specifications for the data highway system were reviewed and are ready to be ordered. It has been decided that the hardware installation and the software programming will proceed in parallel only if adequate resources for technical support is available. Otherwise, the hardware installation will take precedence to programming in order to maximize the learning advantage.

DISCUSSION AND CONCLUSION

Unlike most manufacturing undertakings which center around investment decisions, namely ROI, this experiment represents more a policy decision. There is no ROI justification; it is therefore a matter of conviction rather than accounting. The hope is to design a prototype which scrutinizes the feasibility of arriving at CIM by way of stand-alone equipment. The acquired knowledge or experience can and will benefit other manufacturing operations which share similar needs and possess a similar environment. The conclusion derived from this specific case seems to indicate that for companies with high existing capital equipment or facility, a sound approach will be to integrate this equipment by providing the missing links with new additions.

A new dimension of this experimental case is its representation of the joint effort between industry and academic institutions. Feedback from the CIDMAC participants explicitly stated that the experience was educational as well as economically beneficial (to the Company). A great deal of insight was acquired on what is actually involved in software programming for a "real-world" application, as compared to "test-lab" type of programming. Such joint effort representation serves as a stimulant to further industrial-academic involvement.

The principle of capital effectiveness was demonstrated in this experiment. It showed that CIM does not automatically mean expensiveness, provided sufficient planning is done at the project design stage. The capital expenditure of this experiment provides sufficient evidence to support the fact that from stand-alone to CIM can be accomplished economically. The following observations based on experimental results will serve as good references for groups who are interested in pursuing the CIM cell approach:

- A large amount of emphasis should be placed on project management, and progress should be monitored closely. At the outset, scope and objectives must be clearly mapped out jointly by engineering and operation groups. During the life cycle of the project, technical details

(continued)

must be worked out and documented to ensure software/hardware compatibility. Continuous and consistent technical support is crucial; every effort must be made to avoid chancing baton at mid-stream.

- Avoid replacing hardware by software. It might look attractive economically to do so, in order to reduce up front or initial costs. However, in-house software development is very time-consuming, and the quality of performance is dubious. The experience of this project indicates that hardware specification and selection should preceed software development.

- If PCs are used, control and sequencing of machine operations are best handled by ladder logic, which is specifically designed for such purposes.

- For data transfer and information processing, the recommendation is to use higher-level languages such as BASIC or FORTRAN. Ladder logic is capable of handling the task, but it appears to be too intricate for trouble shooting and somewhat clumsy for efficient programming. Most PC suppliers now have intelligent modules that fit into existing I/O racks. These modules have the capabilities of a high-level language.

- Effort should be directed toward making the program user friendly and "dummy-proof" to accommodate shop floor environment and to enhance man/machine interface.

- When data are needed for information processing reasons, software can be greatly simplified by having the master controller read the memories of other controllers on the network.

- Avoid chasing technology. New products hit the market every six months and will continue to do so. In designing a CIM Project, the project engineer must "freeze" the technology content at a given time to allow the project to proceed. Caution needs to be exercised to ensure the current design can be adapted to future technology.

- It was pointed out that computerized integration impacts the shop floor human resources and structure differently from stand-alone environment. Consideration needs to be emphasized on job design, training, and organizational issues.

Overall, this experimental case appeared to be very educational, and experience thus gained will definitely be very valuable for future projects. Besides demonstrating the feasibility of moving toward CIM from stand-alone environment, this project presented a message which says the term "TURNKEY" needs to be clearly defined. In automation or computerized integration, there is no substitute for the actual involvement by the users. A project team must be in place to launch any integration effort. The size of the team may vary according to the project magnitude, but the truth remains that the word "TURNKEY" is elusive, and the success of any of such system hinges entirely on the joint design effort between the user and the supplier.

The future of CAD/CAM can be summarized very simply and succinctly. It is the completely integrated, completely automated manufacturing plant. In such a plant, the traditional walls between design and manufacturing will come down. The database developed in designing products will be shared directly with manufacturing. All design and manufacturing components will be integrated into one large system.

However, before this future development becomes a reality, there will be many other developments in the various individual CAD/CAM areas of CADD, CNC, robotics, and CIM.

FUTURE DEVELOPMENTS IN CADD

There are a number of technological developments which can be expected in this area. Developments in microelectronics will allow more and more memory to be built into computers which will, in turn, increase the number of drawings and other forms of documentation that can be stored in the main memory of a computer. This, coupled with continued development in software will make three-dimensional solid modeling the norm in future design settings. Technological advances in the area of color graphics and plotter technology will increase significantly the amount of color documentation used. The use of database management software will also become common in CADD. However, the single most significant future development in this area will be the rapid growth of *microCADD*.

By the turn of the century, design and drafting on microcomputer-based systems will be the norm in design and drafting. The transition away from the larger traditional mainframe and minicomputer-based systems is already well underway. As microcomputer technology continues to improve, the rate of the transition will increase correspondingly. Microcomputers already have the memory, processing speed, and graphic capabilities needed to accomplish even the most difficult CADD tasks, such as solid modeling. As the technology continues to develop, they will be even faster at performing such tasks. The one remaining missing element is the fully developed, broadly compatible communications networking capability for communication among stations and other access points.

MicroCADD began to be seen in 1979 and 1980. At that time, there were only three microCADD vendors. By 1985, this number had grown to over 160. The microCADD industry is experiencing over 60% growth in its annual revenues, compared with 37% for traditional CADD vendors. In 1990, there will be almost 900,000 microCADD systems in use in this country, compared with approximately 500,000 traditional systems. By the year 2000, the typical design and drafting station will be a high-powered, high-resolution, color-capable microcomputer with the necessary peripherals.

FUTURE DEVELOPMENTS IN CNC

The use of CNC machines will increase in the future as controllers become more and more capable but less and less expensive. The inclusion of CNC machines in machine cells and CIM systems will also contribute to broader acceptance and application of CNC.

The three most significant areas of development with regard to the future of CNC have to do with where the machines will be used. They will be used increasingly in

1. smaller shops and manufacturing firms,
2. machining cells, and
3. CIM systems.

Small Shops and CNC

Typically large manufacturing firms have led the way in converting from manually controlled machines and systems to CNC. For small shops, CNC has been something to wish for. This has been due, principally, to the expensive initial costs. However, the future will see this inhibitor removed.

As more and more vendors enter the CNC market, competition will lead to better products and lower prices. This is especially true with regard to programmable controllers. This device will follow the pattern set by the computer of smaller, more powerful, but less expensive models turned out every year. Each successive model will tend to be simpler to operate and easier to learn. These factors will lead to wider use of CNC machines and systems in small shops and manufacturing firms.

CNC and CIM Systems/Machining Cells

One of the main components of a CIM system is the machine tool component. Machine tools in CIM systems are CNC machines. As manufacturing technology inches closer to the realization of CIM, more and more CNC machines will be used in CIM systems.

Large CIM systems are sometimes collections of smaller machining cells. A machining cell is one or more CNC machines and some type of parts handling device or system. Since the main ingredient in a machining center is the CNC machine, the expanded use of machining centers will result in a future broadening of the use of CNC machines and systems.

FUTURE DEVELOPMENTS IN ROBOTICS

The use of industrial robots in manufacturing will increase rapidly in the years to come, until, by the year 2000, the robot will be a normal part of most larger manufacturing operations. Robotics enjoyed only slow growth in this country during its formative years. In 1965, there were only about 2000 robots in use. By 1985, this number had grown to 15,000. By 1990, it will increase to over 100,000.

In addition to seeing more robots in the future, manufacturing will see better robots. Improvements in power technology will allow for more electrically and pneumatically driven robots. Improvements in control and end-of-arm tooling technology will allow for wider applications of robots in such critical areas as assembly.

However, the most significant future developments in the field of robotics will occur in the area of robot sensing and vision. These developments hold the key to the complete acceptance of industrial robots as a fundamental part of modern manufacturing plants. All of the non-contact sensor technologies will continue to improve. These include proximity, photoelectric, and vision sensors. Vision systems will eventually become the norm with regard to non-contact sensing. By the year 2000, advances in artificial intelligence, microelectronic circuitry, display graphics, and camera technology will allow for the development of industrial robots that can approximate the basic capabilities of human vision.

FUTURE DEVELOPMENTS IN CIM

CIM really is the future of CAD/CAM. In fact, at a point in the future, the term CAD/CAM will be replaced by CIM. At this point, there are CIM systems in which CNC machines, materials handling systems, and robots are grouped in an integrated manner. These are CIM systems, but they are limited systems. The future will see CIM systems in which all components—CAD, CNC machines, materials handling systems, and robots—will be wholly integrated. Such systems will be the key components of factories of the future.

SUMMARY

The future of CAD/CAM can be summarized in one short sentence. It is the completely integrated, completely automated manufacturing plant. Before this future development is realized, however, there will be a number of lesser, but important, developments in the individual CAD/CAM areas.

The most important development in the area of CADD will be the broad-based acceptance of microCADD and its replacement of traditional mainframe and minicomputer-based systems as the norm in design and drafting. CNC will gain wider acceptance as lower costs make it a feasible alternative for small shops and manufacturing plants. In addition, the inclusion of CNC machines in machining cells and CIM systems will increase the use of CNC.

The number of industrial robots installed will increase significantly, such that by 1990 there will be over 100,000 robots installed in this country alone. In addition to more robots, the future will see better, more capable robots as technological advances in the areas of control, power, and sensing/vision continue to occur.

Today's CIM systems have integrated CNC, materials handling, and robotics. Tomorrow's systems will integrate all CAD/CAM concepts. In fact, the term CAD/CAM will eventually become obsolete and be replaced by CIM.

Chapter Ten REVIEW

1. Explain what the long-term future of CAD/CAM is likely to be.
2. List and explain the most significant future developments to be expected in CADD.
3. List and explain the most significant future developments to be expected in CNC.
4. List and explain the most significant future developments to be expected in robotics.
5. List and explain the most significant future developments to be expected in CIM.

Glossary

Absolute system—A type of numerical control system in which all coordinates are dimensioned and programmed from a fixed or absolute zero point.

Absolute vector—A vector whose endpoints are defined in terms of units from the specified origin.

Access time—The length of time between the instant when a single unit of data is called from memory, and the moment when transmission is completed from the calling device.

Actuator—In robots, a device which converts electrical, hydraulic, or pneumatic energy into motion, i.e., cylinders, servo motors, rotary actuators.

Adaptive control—Machine control units for which fixed speeds and feeds are determined by feedback sensors rather than being programmed.

Address—The location in computer's memory of a "word" or "block" of data.

Addressability—The range of addressable points or device coordinates.

Addressable point—Any position specified in device coordinates.

Algorithm—A step-by-step procedure for achieving a given result by proceeding in a logical manner. CNC and computer programs are developed in this way.

Aliasing—The visual effects that occur when the detail of an image exceeds the resolution of the device space, i.e., a stairstep line on a raster display.

Allowed time—The leveled time with allowances for fatigue and delays added on.

Alphanumeric display—A CRT display used to display text strings.

Analog computer—A computer working on the basis of a physical analogy (comparison of similarities) of the mathematical problem to be solved. The computer translates temperature, speed, voltage, or other physical variables into related electrical quantities and uses electrical equivalent circuits as an analog.

Annotation—The presence of textual descriptions on a display.

Area fill processor—See **Fill**.

Artificial Intelligence (AI)—Built-in capability of a machine to improve its own operations. The capability of a machine to perform functions that are normally associated with human intelligence, such as learning, adapting, reasoning, self-correction, and automatic improvement.

ASCII—(Pronounced "askey") Abbreviation for "American Standard Code for Information Interchange." ASCII is the standard code for expressing numbers, letters, and a variety of common typewriter symbols, such as period, comma, question mark, and carriage return, in a seven-bit format.

Assembly drawing—A CAD/CAM display which represents a major subdivision of a final product.

Assembly language—The most elementary language in which a human can program a computer.

Associative dimensioning—The updating of the respective dimensions of CAD/CAM display groups as the dimensions of their display entities change.

Asynchronous—Transmission of information between computers is under control of interlocking (to and from) control signals rather than a direct function of clock cycles within a system.

Attribute—Any characteristic of a display item (color, linestyle, character font, etc.) or segment (visibility, detectability, etc.).

Automated assembly—Assembly by means of operations performed automatically by machines. A computer system normally monitors the production and quality levels of the assembly operations.

Automated Materials Handling (AMH)—Handling and moving of materials using such automated devices as robots.

Automated Process Planning (APP)—Creation, with the assistance of a computer, of a process plan for parts in a given family.

Automated Storage and Retrieval (ASR)—The use of a computer for the storage and retrieval of manufacturing and management data.

Automatic Parts Programming System (APPS)—System which allows an operator to manually trace a two-dimensional part for the purpose of producing an NC part program.

Automatic programming tools—The combination of the APT language, the APT computer program and the computer used for implementing the program.

Automation—Converting a procedure, process, equipment, or system to automatic operation.

Axis—A rotary or translational (sliding) joint in the robot. Also called **degree of freedom.**

B-spline—A mathematical representation of an arbitrary smooth curve.

Back annotation—The extracting of information from a completed printed-circuit board to create a CAD/CAM display.

Backed-up—Work that is copied onto a tape or disk to protect against losing data if the "user" tape or disk is damaged.

Batch processor—A computer which accepts programs in a continuous stream and executes them one-by-one, analogous to a batch manufacturing process.

Baud rate—The rate, in bits per second, that data can be transmitted over a serial transmission line, such as a telephone line.

Benchmark—Any set of standards designed to verify performance specifications or to compare hardware or software produced by different manufacturers.

Binary code—Any system of representing data with zeros and ones only.

Bit—A single digit, 0 or 1, of a binary code.

Bit plane—The hardware used as a storage medium for image bit maps.

Blending—Mixing two or more lots of material to produce a homogeneous lot. Blends normally receive new identification and require retesting.

Block—A fundamental unit of data on a disk or tape drive. A commonly used block size is 1,024 bytes, although other block sizes may be defined by the computer system designer.

Block diagram—A graphical presentation of the paths along which data flows between various parts of a computer system.

Bomb-out—The complete failure of a computer routine, requiring a restart or reprogramming.

Boxing—A visibility test incorporated in clipping which uses a bounding box to test the relationship of an entire symbol to the clipping boundaries.

Buffer—Temporary memory, which speeds the flow of information between parts of a computer system by eliminating delays.

Bug—A minor computer malfunction usually caused by a programming error.

Bulk storage—Memory units (separate from the main memory of a computer) in which large amounts of data may be stored.

Button device—A button used as a graphic input device.

Byte—A group of eight bits; a common measure of computer memory capacity.

CAD—Computer-Aided Design. Using computers to aid or enhance the design process.

CADD—Computer-Aided Design and Drafting. Using computers to aid the design process and in producing the documentation of a design.

CAD/CAM—Computer-Aided Design/Computer-Aided Manufacturing. Using a computer to improve productivity in design and manufacturing. Includes CAD, CADD, CNC, robotics, and CIM.

CAM—Computer-Aided Manufacturing. The use of computers to aid in any or all phases of manufacturing.

Cathode-Ray Tube (CRT)—An electron tube in which electron beams projected onto its display surface excite the phosphor coating, producing luminous spots.

Central Processing Unit (CPU)—The heart/brain of a computer, containing all the circuitry for controlling, timing, performing arithmetic, and addressing data for storage or transmission.

Character—An instance of a numeral, letter, or other linguistic, mathematical, or logical symbol.

Character font—A primitive attribute of text strings defining the style of the character set.

Character generator—A hardware device which accesses character patterns in a ROM and generates them at user-specified display surface positions.

Character plane—A primitive attribute of text strings defining the plane in which characters are generated.

Character size—A primitive attribute of text strings defining the size of characters in terms of the bounding box.

Chip—A very small silicon wafer containing an integrated circuit.

Clipping—The process of determining which portion or portions of a display element lie outside the specified clip boundary and making them visible.

CNC—Computer Numerical Control. Control of manufacturing machines and systems using a computer and programmed instructions.

Coherence—A property used in raster scan which recognizes that adjacent pixels are likely to be similar in characteristics.

COM—Computer output microfilm.

COM recorder—A display device for placing displays on microfilm.

Comparator—A device which compares the proximity of a cursor to the vector currently being drawn.

Compatibility—The degree to which tapes, languages, programs, machines, and systems can be interchanged, or can communicate.

Component—A CAD/CAM marker which has physical meaning, i.e., resistor, capacitor, switch.

Composite color—A color described in terms of its hue, whiteness, and blackness, and encoded in a single video signal.

Composite video—A single video signal encoding RGB data.

Computer—An electronic device that manipulates and processes data according to programmed instructions. Basic computers consist of input/output devices, a central processing unit, and memory devices.

Computer-Aided Design (CAD)—The application of computers to aid in any or all phases of the design process.

Computer-Aided Engineering (CAE)—Use of a computer, special software, and various peripheral devices to accomplish such engineering tasks as analysis and modeling.

Computer-Aided Manufacturing (CAM)—Utilization of the computer in the management, control, and operations of the manufacturing through either direct or indirect interface with the physical and human resources of the organization.

Computer animation—The use of computer graphics to generate motion pictures.

Computer Automated Process Planning (CAPP)—Using a computer to develop a detailed plan for the production of a part or assembly.

Computer graphics—A family of related technologies which permit digital computers to control, alter, and display pictures rather than only text or numbers.

Computer Integrated Manufacturing (CIM)—Computer integrated manufacturing is the total integration of such individual concepts as CAD, CNC, robotics, and materials handling into one large system.

Configuration—A group of machines, devices, parts, and other hardware which make up a system.

Connectivity—Data which describe how components of a system are connected.

Continuous path—A servo-driven robot that provides absolute control along an entire path of arm motion, but with certain restrictions in regard to editing and ease of program change.

Contrast—The ratio of the highest available intensity level to the lowest.

Controlled path—A servo-driven robot with a control system which specifies or commands the location and orientation of all robot axes.

Coordinate—The location of a point in terms of units from the specified origin.

Core memory—An old-fashioned term for the main memory or random access memory (RAM) of a computer.

Core system—A proposed graphics standard developed by the ACM Special Interest Group on Graphics (SIGGRAPH).

Crash—A computer system failure which usually results in loss of data and, therefore, loss of labor hours.

Critical items—Production items which require a lead time longer than the normal planning span time, or items whose scarcity could limit production.

Critical path method (CPM)—A special technique for scheduling resources to accomplish a job within all applicable constraints.

Cross hairs—Two intersecting perpendicular lines incorporated in a cursor, with the intersect being used to indicate desired device coordinates.

CRT display—A display device employing a cathode-ray tube.

Cursor—A recognizable display entity that can be moved about the display surface by a graphic input device to return either device coordinates or a pick stack.

Cybernetics—The science of control and communication systems. It encompasses (a) integration of communication, control, and systems theories; (b) development of systems engineering technology; and (c) practical applications at both the hardware and software levels.

Data processing—The performance of a systematical sequence of mathematical and/or logical operations that a computer performs on data.

Database—Any collection of information having predetermined and useful organization.

DBMS—Data Base Management System. A software system for managing data and making such features as interrogation, maintenance, and analysis of data available to users.

Debug—Troubleshooting and correcting computer hardware or software problems that affect the operation or performance of the system or a program.

Decluttering—The selective erasure of display items when the display is too dense to easily discern details.

Delphi method—A forecasting method in which the opinions of experts are combined in a series of questionnaires. The results of each questionnaire are used to design the next questionnaire, so that convergence of the expert opinion is obtained.

Depreciation—The actual decline in the value of an asset due to exhaustion, wear and tear, or obsolescence; such as the depreciation of a piece of equipment.

Digital computer—As contrasted with an analog computer, which operates on continuous data, a digital computer solves problems through arithmetic operations on numbers composed of digits.

Digital vector generator—A device used with raster displays to interpolate the straightest possible pixel string between specified endpoints.

Digitize—Convert from graphic or analog form to representation by discrete values.

Digitizer—A data tablet that generates coordinate data from visual data through the use of a puck or stylus.

Dimensioning—The measuring of distances on a CAD/CAM display.

Direct Numerical Control (DNC)—The use of a shared central computer for distribution of part programs and data to several remote machine tools.

Direct-view storage tube—A type of CRT whose display is maintained by a continuous flood of electrons.

Directed beam—The technique used in calligraphic displays to produce vectors by having the electron beam stroke them in a selected order.

Directed Numeric Control (DNC)—Machines controlled by a dedicated computer that stores numerous N/C machine programs.

Disk—Flat, circular, magnetic recording device capable of storing large amounts of data.

Disk Operating Systems (DOS)—A program which interacts with the processor and the disk or diskette drive to control the flow of data.

Diskette—Sometimes called a "flexible disk" or "floppy disk." A very small, very inexpensive form of disk drive.

Display—A collection of display items presented on the display surface.

Display device—An output device used to display computer-generated graphical data.

Display entity—A logical grouping of output primitives which forms a recognizable unit on the display surface.

Display file—A collection of display instructions assembled to create a display.

Display image—The portion of an image visible on the display surface at any one time.

Division of work—The separation of tasks into less complex subtasks. The basis for work specialization.

Dot matrix—A pattern of dots taken from a two-dimensional array.

Dot matrix plotter—A plotter which produces displays in dot matrix form.

Dragging—The interactive mode technique of moving a display item by translating it along a path determined by a graphic input device.

Draw—The generation of a vector by creating a line segment from the current position to a specified endpoint, which becomes the new current position.

DRC—Design rules checking.

Drum plotter—A plotter whose display surface is a rotatable drum and whose plotting head can only move parallel to the drum's axis of rotation.

Dumb terminal—A terminal which receives and displays data but is incapable of altering it.

DVST—Direct-view storage tube.

Echo—The mode of a graphic input device which provides visual feedback to the operator, e.g., a cursor, text strings, etc.

Electron gun—The part of a CRT which focuses and emits the electron beam.

Electrostatic plotter—A raster plotter which produces display images on paper sensitized to electrostatic charges.

Element—A basic graphical entity, i.e., a point, line segment, character, marker, or text string.

Emulation—The use of special programming techniques and machine features to allow a computing system to execute programs written for another system.

End effector—The tool attached to the robot manipulator or arm that actually performs the work.

Endpoint—Either of the points that mark the ends of a line segment.

Endpoint matching—The accuracy of the vector generator in drawing two or more vectors emanating from the same point.

Expansion card—A card on which additional printed circuits can be mounted.

Expediting—The "rushing" or facilitating of production orders which are needed in less than the normal lead time.

Fabrication—Processing of materials for desired modification of shape and properties.

Family of parts—A group of parts having similar topology or manufacturing characteristics.

FEM—Finite-element model.

Field—A part of a data record set aside for a specific element of data (see "database").

File—A collection of data, in the form of records, that exists on a computer's disk or tape.

Fill—To fill an area of the display surface bounded by vectors, e.g., with a solid color or a pattern of line segments.

Finite element analysis—A method for detailed stress analysis of irregularly shaped structures, such as automobile or airplane frames, nuclear reactor nozzles, etc.

Finite element model—A mathematical model of a continuous object which divides the object into an array of discrete elements for the purpose of simulated structural analysis.

Firmware—Sets of computer instructions (programs) cast into Read Only Memory.

Fixing—The positioning of a display item at a set location after dragging.

Fixture—A device needed to hold a workpiece in proper position for work performance.

Flatbed plotter—A plotter with a flat display surface fully accessible by the plotting head.

Flexible automation—Refers to the multitask capability of robots; multi-purpose; adaptable; reprogrammable.

Flicker—A noticeable flashing of the display during each refresh, caused when the refresh interval exceeds the phosphor persistence.

Floating point arithmetic—Arithmetic performed with real numbers, such as 12.5, 3.1417, and 87.66, as distinguished from integers, such as 1, 753, and 56.

Floppy disk—A storage device consisting of a flexible magnetic disk inside a protective plastic jacket.

Flowchart—A diagram that outlines the logical steps a computer program should take, facilitating program design or documentation.

Fonts—Line fonts are repetitive patterns used to give meaning to a line such as hidden lines, centerlines, or phantom lines.

Function button—A button on a button device which can operate in either momentary or latchable mode, and whose value may be retained.

Function key—A key on a function pad which causes execution of special program functions defined by the user.

Function pad—A graphic input device with user-programmable function keys.

Function switch—A button on a button device which can operate in either momentary or latchable mode, and whose value may be retained.

GNC (Geographical Numerical Control)—A part programming system which provides effective tape generation by providing graphic displays of the part, the tool path, and the tools themselves.

Graphic input—Any inputs entered by a user through a graphic input device while in interactive mode.

Graphic input device—The hardware which allows the user to enter data, or pick a detectable display item.

Graphics—The visual presentation of data as a series of output primitive.

Graphics package—A series of software routines which provide the user access to the graphics hardware for the purpose of generating a display.

Graphics processor—A controller which accesses the display list, interprets the display instructions, and passes coordinates to the vector generator.

Graphics tablet—A peripheral device for "drawing" on a computer graphics system.

Gravity feed—Supplying materials into a machine, workstation, or system by the force of gravity.

Grid—The uniformly spaced points in two or three dimensions within which an object may be defined.

Group classification code—Material classification technique which designates characteristics using successively lower order groups of code.

Group technology—A means of classifying into families on the basis of similarities of the parts.

Hard copy—Information printed on paper instead of on the graphics display tube.

Hardware—The assembly of electronic components that constitute the physical makeup of a data processing system.

Hatching—The filling of an area of the display surface bounded by vectors with a pattern of parallel line segments.

Hexadecimal—Numbers written in base 16 rather than base 10; used to display binary data because the hexadecimal form is more compact, yet can be easily translated into binary numbers.

High-resolution graphics—Graphic terminals that show great detail because they are made up of many pixels.

Homogeneous coordinates—The coordinates used in matrix transformations to convert objects described in n-space to a representation described in n + 1 space; i.e., x, y, z become wx, wy, wz, where w is the homogeneous factor.

Host computer—The "independent" computer to which a peripheral device, such as a terminal, plotter, or disk drive, is attached.

Hue—A characteristic of color which allows it to be named (i.e., red, yellow, green, blue) and which is often defined by an angle representing its graduation.

IGES—Abbreviation for "Initial Graphic Exchange Standard"; a U.S. national standard for exchanging mechanical design data between CAD systems.

I/O device (input/output)—A device, such as a disk drive, which can send and receive data from a computer.

Image—A view of an object.

Image bit map—A digital representation of a display image as a pattern of bits, where each bit maps to one or more pixels.

Image transformations—The applying of a transformation function to an image after projection to the display area.

Indirect labor—Labor which does not add to the value of a product but must be performed as part of its manufacture.

Ink-jet plotter—A plotter which uses electrostatic technology to first atomize a liquid ink and then control the number of droplets that are deposited on the plotting medium.

Input—Data (or programs) which go into a computer system.

Input device—Any device such as a digitizer or keyboard that allows the user to give information to the computer.

Inspection by attributes—Inspection in which a part is accepted or rejected based on a single requirement or set of requirements.

Integrated circuit (IC)—A microelectronic chip.

Intelligent robot—A robot which can control its behavior through the application of its sensing and recognition capabilities.

Intensity—A characteristic of color defining its percentage ranking on a scale from dark to light, specifying perceived brightness.

Intensity level—One of a discrete set of brightness levels attainable with a CRT.

Interactive graphics—A method which allows users to dynamically modify displays through the use of graphic input data.

Interactive processor—A "personal" computer, which performs the user's instructions immediately upon receiving them, and displays the results immediately upon completion.

Job classification—The grouping of jobs on the basis of the functions performed, level of pay, job evaluation, historic groupings, collective bargaining, or any other criteria.

Job lot—A small number of a specific type of part or product that is produced at one time.

Job lot layout—A group of machines and equipment especially arranged to handle job lot production.

Joystick—A graphic input device which employs a movable lever to control the position of a cursor, for returning locator or pick information.

Kernel—A subset of routines from a graphics package which permits construction of elementary displays.

Keyboard device—A graphic input device which allows the user to enter characters or other key-driven values.

Labor cost—That part of the cost of a product attributable to wages.

Laser plotter—A plotter which produces display images on photographic film, in raster or vector formats, using a laser.

Layer—The logical subdivisions of the data contained in a two-dimensional CAD/CAM display, such that the subdivisions may be viewed individually or overlaid and viewed in groups.

Lead time—The time normally required to perform a given activity.

Library—Usually a collection of predefined symbols which may be placed on a CAD system drawing.

Light pen—A graphic input device which generates a hit detection when a pick is made while pointed at a detectable display item.

Limit switch—A switch that is actuated by some motion of a machine to alter the electrical circuit and limit or guide movement.

Line style—A primitive attribute of lines which defines whether they are to be solid or dashed, and a possible dash pattern.

Line type—See line style.

Line width—A primitive attribute which defines the thickness of a line segment.

Local area network (LAN)—A type of network hardware which doesn't allow computers to be separated by physical distances of more than a thousand yards.

Locator device—A graphic input device, such as a joystick or data tablet, which uses a cursor to provide coordinate information.

Logic diagram—A drawing which indicates the interconnection of the individual logic elements of an electronic circuit.

M Function—Similar to a G function except that the M functions control miscellaneous functions of the machine tool, such as turning on and off coolant or operating clamps.

Machine center—A group of similar machines which can all be grouped together for purposes of loading. See **Work Center**.

Machine tool—A powered machine used to form a part, typically by the action of a tool moving in relation to the workpiece to perform such tasks as turning, drilling, cutting, etc.

Macro—A sequence of CAD system commands contained in a text file or grouped on a menu key, which may be executed automatically.

Magnetic tape—Plastic or mylar tape that is coated with magnetic material; used to store information.

Magnetic tape storage—The recording of binary data on a magnetized tape.

Mainframe—An overused, ill-defined term which refers to the central processing unit(s) of a large data processing system.

Maintenance—The collection of procedures necessary for retaining an item in or restoring it to a specified condition.

Manipulator—The "arm" of the robot. Encompasses mechanical movement from the robot base through the wrist.

Manufacturing—A series of interrelated activities and operations involving the design, material selection, planning, production, quality assurance, management, and marketing of goods and products.

Manufacturing Planning and Control Systems (MP&CS)—Special systems, usually automated, used to set the limits or levels of manufacturing operations in the future and to control machines, materials, and processes in the present.

Manufacturing process—A series of tasks performed upon material to convert it from the raw or semifinished state to a state of further completion and a greater value.

Mapping function—A method of transforming an image definition expressed in one coordinate system to another.

Mass production—A method of high quantity production characterized by a high degree of planning, specialization of equipment and labor, and integrated utilization of all productive factors.

Materials handling—The movement of materials, parts, subassemblies, or assemblies either manually or through the use of powered equipment, robots, or systems.

Material Requirements Planning (MRP)—Process for identifying the types and amounts of materials required for future manufacturing projects using bills of material, inventories, and master production schedules.

Memory capacity—The number of actions that a robot can perform in a program.

Menu—A list of program execution options appearing on the display surface which prompts the user to choose one or more through the use of a graphic input device.

Menu keypad—An electronic tablet that contains individual blocks that may be programmed with single commands or groups of commands.

Microprocessor—A central processing unit contained on a single integrated circuit.

Mirroring—The creation of a mirror image of a display image.

Modeling system—A system which allows models to be defined and transformed using world coordinates.

Modeling transformation—A transformation which transforms the world coordinate system of a model to the default world coordinate system of a graphics package which is in effect immediately prior to a viewing operation.

Modem—Short for "modulator-demodulator," and permits a digital computer to transmit digital data over the analog circuits of local telephone lines.

Motion study—An analysis of the movements which occur in an operation for the purpose of eliminating wasted movement and establishing a better sequence.

Mouse—A device for controlling the cursor. It looks like a small box, with several buttons on the top, and is connected to the computer by a cable.

Nesting—In programming, the grouping of individual commands or operations into a single command that contains all of the various operations.

Networks—Communication lines connecting computers and computer-controlled systems.

Node—The intersection of two or more interconnections.

Numerical control—A method of controlling manufacturing equipment and systems which accepts commands, data, and instructions in symbolic form as an input and converts this information into a physical output in operating machines.

Off-line—Equipment or devices in a data processing system which are not under the direct and immediate control of the central processing unit.

On-line—Equipment or devices in a data processing system which are under the direct and immediate control of the central processing unit.

Open shop—A CAD facility in which designers from a variety of departments use design stations on a part-time basis.

Operating system—The primary control program of a data processing system. Written in assembly language, the program controls the execution of programs within the CPU and controls the flow of data to and from the memory, and to and from all peripheral devices in the data processing system.

Optical Scanner—A video camera tube incorporating an electron beam to scan an input image, sense the light emitted, and produce video signals.

Optimize—The arrangement of instructions in numerical control or computer applications to obtain the best balance between operating efficiency and the use of hardware capacity.

Output—Data which come out of a computer.

Painting—A technique similar to inking, but used only on raster displays where line width and color may vary.

Pan—To translate horizontally.

Paper tape—One type of input medium used in numerical control.

Parallax—The apparent displacement of a display item from where the viewer perceives it and where a light pen is pointing.

Parallel operation—The performance of several actions (usually of a similar nature) simultaneously.

Parallel transmission—A system of transmitting information wherein the characters of a word are transmitted simultaneously over separate lines or wires (as opposed to serial transmission).

Part—A single manufactured item used as a component in an assembly or subassembly.

Parts list—List of parts used in an assembly or subassembly.

Passive graphics—A method allowing no operator dynamic interaction with a display.

Passive mode—A setting which specifies a display console as usable for passive graphics.

Pattern fill—Repetitively using a user-defined pixel array to perform fill.

Payload—Maximum weight carried at normal speed.

Pen plotter—A vector which draws with ordinary ballpoint pens and ink.

Peripheral (device)—Any device distinct from the central processing unit, such as disk drives, CRTs, plotters, and graphic tablets.

Phosphor—The coating of the inside of a CRT.

Pick-and-place robot—A simple robot with usually 2-4 axes of motion and little or no trajectory control.

Pixel—The discrete display element of a raster display, represented as a single point with a specified color or intensity level.

Pixel array—A rectangular matrix of pixels.

Plasma panel—A type of display device whose display surface consists of a matrix of gas-filled cells which can be turned on and off individually and which remain on until turned off.

Plotter—Any device that produces hardcopy of graphic data. Types of plotters include vector (pen) plotters, electrostatic plotters, and ink-jet plotters.

Point-to-point—A servo or non-servo-driven robot with a control system for programming a series of points without regard for coordination of axes.

Pop-up menus—Instead of having to memorize hundreds of commands, the program permits standard system commands to appear, as needed, on little "cards" which float on the screen.

Ports—A physical connection linking a processor to another device or circuit.

Printed circuit board—A board on which a pattern of printed traces and connections has been etched.

Process—Series of continuous actions with a system of levels divided into subactivities that are accomplished by executing one or more tasks.

Production—Changing the shape, composition, or combination of materials, parts, or subassemblies in manufacturing.

Production capacity—The highest sustainable output rate which can be maintained without changing the product specifications, product mix, worker effort, plant, and equipment.

Production rates—The quantity of production expressed in a measurable unit such as hours, or some other broad measure, expressed by a period of time, i.e., per hour, per shift, per day, per week, etc. (Same as Production levels.)

Productivity—The output of goods measured against some standard, norm, or potential maximum.

Program—A set of instructions written in a data processing language which defines a task to be performed by a computer.

Prompt—Any action of the display console which indicates an operator reaction is needed.

Protocol—A set of rules governing the exchange of data between devices in a data processing system.

Puck—A handheld device with a transparent portion containing cross hairs that is used for inputting coordinate data from a data tablet through the use of programmable buttons.

Quality assurance (QA)—A broad term which includes both quality control and quality engineering.

Quality control—The establishment of acceptable limits in size, weight, finish, and so forth for products or services and of the resulting goods or services within these limits, through inspection, testing, gaging, and so on.

Random Access Memory (RAM)—An array of semiconductor devices for temporary storage of data during the computation process.

Raster—A rectangular matrix of pixels.

Raster display—A CRT display whose display surface is covered by a raster and which generates displays using raster scan techniques.

Raster plotter—A plotter which produces displays in dot matrix form.

Raw material—Unprocessed material.

Read—To query a graphic input device and await operator action.

Real-time—Computer processing or control performed at the same time as the controlled operation or process is occurring. Also, accessing data that represent actual values at that point in time.

Refresh—The process of repeatedly drawing a display on the display surface of a refresh tube.

Refresh cycle—One refresh of the display surface.

Refresh display—A display device employing a refresh tube which permits dynamics because of high refresh rate.

Register—A circuit that stores bits in a CPU.

Repaint—To refresh a display surface with an updated display.

Repeatability—The accuracy of an analog vector generator in minimizing the deviation from precise overlap when redrawing vectors.

Resolution—The precision of a CRT, measured as the number of line pairs distinguishable across the display surface.

Robot—An electromechanical device that performs functions traditionally performed by humans.

ROM (Read Only Memory)—Similar to RAM, except that the computer user may only retrieve data, not change it.

Rotate—To transform a display or display item by revolving it around a specific axis.

Routing—The positioning of interconnections in a CAD/CAM display.

Routing sheet—A form listing the sequence of operations to be used in producing a particular part or product.

Rubber banding—A programming technique which allows lines drawn on a computer graphics display to be stretched as if they were made of rubber bands.

Saturation—A characteristic of color defining its percentage difference from a gray of the same blackness.

Saved—Work that is filed or stored on a disk drive, so that loss of power or machine failure will not destroy it.

Scale—(1) To transform the size or shape of a display or display item by modifying the coordinate dimensions. (2) The ratio of the actual dimensions of a model to the true dimensions of the subject represented.

Scan line—A horizontal line of pixels on a raster display that is swept by the electron beam during refresh.

Scanning pattern—The path followed by an exploring spot.

Scanning spot—The point of focus of the electron beam of an image digitizer on the input image.

Scissoring—The process of determining which portion or portions of a display element will not be visible.

Screen coordinate system—A coordinate system which represents the internal digital limits of the display device.

Scrolling—The translating of text strings or graphics vertically.

Sector—A subdivision of a block, usually the smallest unit of data which can be retrieved from disk or tape memory.

Serial transmission—The transmission of data over a single pathway.

Servo mechanism—An amplifying device which takes an input from a low-energy source and directs an output requiring large quantities of energy.

Servo system—A control linking a system's input and output which provides feedback on system performance to regulate the operation of the system.

Setup lead time—The time needed to prepare or set up a manufacturing process.

Shading—(1) An image processing technique which indicates light sources in a three-dimensional image. (2) The changes in sensitivity of the video camera tube of an image digitizer.

Shielding—The defining of an opaque viewport or window in which to display a menu, a title, or a message to the operator.

Signal—A logical linking of pins in a CAD/CAM display using interconnections.

Simulation—The design and operation of a model of a system, in a manner analogous to the way the real system operates.

Soft copy—A copy of a display in video form, as on videotape.

Software—Programs, procedures, and associated documentation for data processing systems.

Solid modeling—The construction of a "solid" model of a part from mathematically defined solid primitives (i.e., cubes, cones, and spheres) or surface primitives (i.e., planes, spherical segments, and deformed surfaces).

Specification—A statement of the technical requirements of a material, a part, or a service, and of the procedure to be used to verify that the requirements are met.

Stand-alone workstation—A CAD workstation containing microprocessors and capable of independent operation without being connected to a host computer.

Storage—The process or location for holding data or instructions inside a memory device.

Storage tube—A CRT which maintains a display on the display surface without refresh.

Stroke writing—A line drawing, as opposed to a raster scan.

Stylus—A device analogous to a pencil which is used in conjunction with a data tablet to input coordinate information.

Subroutine—A named display item description contained in the display list, used to create multiple views of the item without repeating the display instructions.

Synchronous—An operation that takes place within a fixed time interval under the control of a clock.

Syntax—Set of rules dictating acceptable grammar of a computer language.

Tablet—A flat-surfaced graphic input device used with a stylus for inking and cursor movement, or with a puck for digitizing.

Tape drive (magnetic)—The hardware device which records and reads magnetic tape.

Tape punch—A special device which punches holes in a paper tape to record coded instructions.

Tape reader—A special device for reading and interpreting information stored on tape and converting it to electrical signals.

Teach pendant—The control box which an operator uses to guide a robot through the motions of its tasks.

Terminal—A peripheral device for entering data into a data processing system or for retrieving data from it.

Text string—A collection of characters.

Thumbwheel—A graphic input device consisting of a rotatable dial which controls the movement of a line across the display surface, horizontally or vertically.

Timesharing—The sharing of one central processor by several terminals.

Tolerance—An allowable variation in a feature of a part.

Tool center point (TCP)—A given point at the tool level around which the robot is programmed for task performance.

Tool design—That division of mechanical design which specializes in the design of jigs and fixtures.

Tool life—The life expectancy of a tool. It is usually expressed as the number of pieces the tool is expected to make before it wears out, or as the number of hours of use anticipated.

Touch-sensitive display—A display device whose display surface can register physical contact.

Trackball—A graphic input device which employs a mounted rotatable ball to control the position of the cursor, used for producing coordinate data.

Translate—To transform a display item on the display surface by repositioning it to another coordinate location.

Turnkey—Any complete system of hardware, software, and service sold for a single price by one vendor.

Utility program—Any program designed to perform routine housekeeping functions in a computer system, such as copying files, making back-up tapes, removing "wasted" space on disk drives, and managing the flow of data to output devices such as plotters.

Vector—A directed line segment, or a string of related numbers.

Vector plotter—A plotter which draws with ordinary ballpoint pens and ink. The pen is driven over a plotting surface by cables attached to pulleys and servomotors.

View plane—The projection plane used in three-dimensional viewing operations.

View point—The originating point of a field of view.

Viewport—A specified rectangle on the view surface within which a window's contents are displayed.

Virtual Coordinate System—The result of mapping a portion of the world coordinate system to the finite limits of the device space.

Vision—The process of sensing and understanding the environment based on the light level of reflectance of objects, i.e., robot vision.

Voice input device—A graphic input device which accepts and interprets vocal data.

Window—The specified area on the view plane containing the projections to be displayed.

Window clipping—The bounding of a view volume in the "x" and "y" directions by passing projectors through the corners of the window to define its sides.

Wire frame—An image of a three-dimensional object displayed as a series of line segments outlining its surface, including hidden lines.

Word—The basic unit which can be handled in a computer. A 16-bit computer handles two bytes per word.

Work center—A specific production facility or group of machines arranged for performing a specific operation or producing a given part family.

Workpiece—A part in any stage of production prior to its becoming a finished part.

Workstation—A configuration containing a display device and any associated graphic input devices.

Wraparound—The positioning of a display item such that it overlaps the border of the device space, resulting in it being displayed on the opposite side of the display surface.

Write protect—A feature which prevents the updating of a bit plane.

Yaw—Side-to-side motion at an axis.

Yon plane—The back clipping plane used in "z" clipping to define a finite view volume.

Zoom—To scale a display or display item so it appears to either approach or recede from the viewer.

INDEX